DEVELOPING CONCISE ENGINEERING MANAGEMENT PLANS

by

Paul Nathan

Chartered Professional Engineer (Leadership & Management), Engineers Australia
Professional Member, Asia Pacific Engineers Council;
Professional Member, International Engineers Mobility Forum;

"Step by Step Guidance for Engineers with Sample Contents of a Concise Engineering Management Plan"

COPYRIGHT © 2018 BY PAUL NATHAN
C/O GUIDESAFE AUSTRALIA
ABN 37899405669
ALL RIGHTS RESERVED. THIS BOOK OR ANY PORTION THEREOF
MAY NOT BE REPRODUCED OR USED IN ANY MANNER WHATSOEVER
WITHOUT THE EXPRESS WRITTEN PERMISSION OF THE PUBLISHER
EXCEPT FOR THE USE OF BRIEF QUOTATIONS IN A BOOK REVIEW.

ABOUT THE AUTHOR

Paul Nathan is a highly experienced Chartered Engineer with Engineers Australia in the Practice Areas of Leadership & Management and Mechanical Engineering.

He has worked in both management and senior management capacities in affiliation with various major blue-chip companies such as ExxonMobil, Shell, <Consultant Name>, Rio Tinto, KONE, Samsung and Imperial Chemicals Industries (ICI) over the last 25 years, both in Australia and across Asia, that too in diversified industries such Oil & Gas, Chemicals, Iron Ore and Construction.

His vast experience in developing world class functional execution plans has immensely benefitted and transformed multi-billion-dollar projects in its project delivery processes to become world class.

He is a natural teacher, trainer and a leader who has the passion to convey his rich experience to leaders and managers everywhere given his in depth as well as up-close and personal experience.

He has managed and led highly effective teams across many of the blue-chip organizations listed above, some of which have won national and global corporate awards.

He currently resides in Perth, Western Australia with his wife and two adult sons and works as a Senior Management Consultant in affiliation with a private consulting organization.

He is available to extend and share his experience in mentoring, coaching and guiding leaders, managers and engineers as well as conducting teaching sessions, seminars and conferences.

Paul can be contacted either via his email address which is pnathan777@gmail.com.

His LinkedIn profile can be accessed via the following link: https://www.linkedin.com/in/paul-nathan-cpeng-mieaust-intpe-apec-engr-16550816/

TABLE OF CONTENTS

I. Introduction.
II. Why the Need for An Engineering Management Plan (EMP)?
III. QA Management of an EMP.
IV. How to Use This Book to Develop an EMP.
V. Guidance on the Proposed "Table of Contents" of a Typical EMP.
VI. Proposed "Table of Contents" of a Typical EMP.

1. Proposed Content for the **"Purpose"** Section.
2. Proposed Content for the **"Scope of Work"** Section.
3. Proposed Content for the **"Terms and Definition"** Section.
4. Proposed Content for the **"References"** Section.
5. Proposed Content for the **"Applicable Procedures, Standards, Specifications and Acts"** Section.
6. Proposed Content for the **"Scope of Engineering Services"** Section.
7. Proposed Content for the **"Organization"** Section.
8. Proposed Content for the **"Work Breakdown Structure"** Section.
9. Proposed Content for the **"Engineering Systems and Information Management"** Section.
10. Proposed Content for the **"Communications and Reporting"** Section.
11. Proposed Content for the **"Schedule Monitoring and Reporting"** Section.
12. Proposed Content for the **"Engineering Controls (Planning and Performance)"** Section.
13. Proposed Content for the **"Engineering Budget and Cost Management"** Section.
14. Proposed Content for the **"Basis of Design"** Section.
15. Proposed Content for the **"Engineering Design Management"** Section.
16. Proposed Content for the **"Health, Safety and Environment (HSE) Management in Design"** Section.
17. Proposed Content for the **"Risk Management"** Section.
18. Proposed Content for the **"Constructability Review"** Section.
19. Proposed Content for the **"Maintainability and Operability Review"** Section.
20. Proposed Content for the **"Modularization and Pre-Assembly"** Section.
21. Proposed Content for the **"Engineering Drawing Management"** Section.

22. Proposed Content for the **"Engineering Coordination with Other Functions"** Section.
23. Proposed Content for the **"Change Management"** Section.
24. Proposed Content for the **"Quality Management"** Section.
25. Proposed Content for the **"Handover and Completions"** Section.
26. Proposed Content for the **"Appendices"** Section.

I.
INTRODUCTION

◆ ◆ ◆

If truth were to be told, many engineering managers and professionals are ACTION ORIENTED and have the tendency to JUMP RIGHT IN and GET THE DETAIL JOB DONE!

I know this for a fact from my own personal experience as well as from my close observations over the years of the engineering professionals whom I have closely worked with.

Further, there is always a time pressure when it comes to delivery of a project which does not help.

There is not much of guidance provided to engineering managers and professionals in making them become aware and appreciate the absolute importance of PLANNING from the onset of the engineering development phase.in making them become aware and appreciate of the absolute importance of PLANNING from the onset of the engineering development phase.

In many engineering practices, the need to expend time and effort in developing an Engineering Management Plan often falls into the "nice to have" category instead of the "must have" category.

Hence, the required resources and time are never allocated for the development of such a plan from the onset and even budgeted as such.

Further, many clients too are not aware of the importance and significance of having such a plan, hence the requirement for such a plan never even gets into the critical deliverables list from the onset of the project initiation stage.

It is no surprise then that the need for such a plan never even appears on the radar of many projects in the first place.

The primary reason for this being the case is because, many engineering managers and professionals are not aware of the significant benefit and advantage such a document brings towards the SUCCESSFUL delivery of a project or an engineering operation.

This book is intended to help and guide engineers, engineering managers and professionals as to the content and framework of a concise Engineering Management Plan.

In the next proceeding section, I have listed the seven (7) primary benefit and value which an Engineering Management Plan provides.

It is my earnest hope and desire that every engineer who reads this book, will gain immense understanding not only as to the typical content of a concise Engineering Management Plan, but MORE IMPORTANTLY of the absolute value in having such a plan in place.

II.
WHY THE NEED FOR AN ENGINEERING MANAGEMENT PLAN (EMP)?

The need for a concise, comprehensive and a practical Engineering Management plan for a given project or engineering related operation cannot be over emphasised.

Such a plan is a crucial **"LIVE"** document that is immense in its value, as listed below:

1. Providing **VISIBILITY** and **TRANSPARENCY** to all stakeholders as to the overall engineering management framework that is adopted which comprises of execution and management methodology, tactics and strategy;

2. Providing **ASSURANCE** to the very stakeholders through this **VISIBILITY** & **TRANSPARENCY** as mentioned above – especially regarding alignment and

adherence to the relevant governing procedures, regulations and applicable laws;

3. Providing an **OPPORTUNITY** for key stakeholders to review and provide their inputs to the engineering management plan framework;
4. Providing a **BASIS** for **QA** to be applied to monitor, assess and evaluate the framework for its effectiveness and relevance;
5. Providing a **BASIS** for **ALIGNING** every member of the Engineering Execution and Delivery Team, which includes sub-contractors, suppliers and partners;
6. Providing a concise and aligned document for the purposes for **INDUCTING & TRAINING** every member of the engineering team.
7. Providing a **BASIS of REFERENCE** and an aligned **FOUNDATION** for the engineering practices.

III. QA MANAGEMENT OF AN EMP

The EMP, being a "live" document, needs to be managed in accordance with the quality and document control management requirements of an organization.
Typically, there needs to be a "Document Owner" assigned which in most cases would either be the Engineering Manager or the Operations Manager, in the absence of the former.

The Document Owner can delegate the responsibility of managing and coordinating the development of this document to either the Quality Assurance Manager or a Senior Engineer with the required experience for the task.

However, the ultimate accountability in finalizing a concise EMP rests with the document owner, who is most often the engineering manager.

As with the QC requirements of many critical documents, the Document Owner will nominate the parties who will need to review and comment on the draft EMP during its developmental stage.

The nomination of the full list of the final stakeholders who will review and sign off on the EMP will need to be determined collaboratively in consultation with the senior managers of the organization and the client representatives, where applicable.

The timeline for the development and issuance of the EMP is solely dependent on obligations undertaken in the contractual agreements.

The veracity and relevancy of the EMP must be QA managed by the Document Owner from time to time.

Some of the common circumstances which require the EMP to be reviewed and updated are as follows:

1. Changes in the Scope & Battery Limits of the Projects/Engineering Operations;

2. Changes in applicable Regulations, Legislations and Procedures (in most cases, "live links" to these are introduced on the EMP to ensure that only the "current" versions are accessed at all time by anyone referring to the EMP);

3. Changes in the Engineering Organization Structure & Assignment of Roles and Responsibilities;

4. Changes related to the pool of key Sub Contractors that are engaged etc.

IV. HOW TO USE THIS BOOK TO DEVELOP AN EMP

In the coming sections, a proposed list of contents that should be typically in an EMP is provided together with proposed write ups for every section of the contents.

These can be exhaustive given that they are based on the requirements of major projects, hence, they can be scaled down and improvised to meet the EMPs for any project.

The proposed list of contents and the proposed write ups provided in this book, **can be copied and/or adopted directly** for the purposes of creating EMPs to suit the specific project requirements.

V. GUIDANCE ON THE PROPOSED "TABLE OF CONTENTS" OF A TYPICAL EMP

Before, we get into the proposed Table of Contents of a typical concise EMP, it is important to note that it is scalable.

What is proposed is typically for a large scale mega project with multi-disciplinary engineering functions working and coordinating across various geographical locations, hence these can be scaled down to meet the scope, type and the battery limits of projects and/or engineering operations.

In developing an EMP, the "audience" for which it is created and the purpose of the EMP must be first and foremost on the mind of the Document Owner, to arrive at a concise and RELEVANT document.

It must be developed with passion and interest with the full appreciation of its value otherwise it will end up becoming a "half-baked and laborious cosmetic document" which will end up on a shelf somewhere collecting dust as it were!

Hence, I urge you approach this exercise of developing a concise EMP with respect, regard and interest.

VI. PROPOSED TABLE OF CONTENTS OF A TYPICAL EMP

Listed below is a comprehensive list of the proposed contents of a typical EMP. Guidance and clarification is provided in the coming chapters on each of these contents, regarding what these respective sections should contain in an EMP.

1. **PURPOSE**
2. **PROJECT OUTLINE**
3. **SCOPE OF WORK**
   - 3.1 Works Excluded
4. **TERMS AND DEFINITIONS**
5. **REFERENCES**
6. **APPLICABLE PROCEDURES, STANDARDS, SPECIFICATIONS AND ACTS**
   - 6.1 Engineering Procedures
   - 6.2 Standards and Specifications
7. **SCOPE OF ENGINEERING SERVICES**
   - 7.1 Disciplines Involved
   - 7.2 Engineering Deliverables
   - 7.3 Engineering List of Work Activities

   7.3.1 Approximate Deliverables for Each Discipline
8 **ORGANISATION**
  8.1 General
  8.2 Locations
   8.2.1 Work Sharing Plan
  8.3 Engineering Organization Structure
   8.3.1 Roles and Responsibilities
   8.3.2 Engineering Manager
   8.3.3 Principal Discipline Engineers
   8.3.4 Lead Discipline Engineers
   8.3.5 Senior Designers
   8.3.6 Drawing Office Manager
   8.3.7 Training and Development
9 **WORK BREAKDOWN STRUCTURE (WBS)**
  9.1 Deviations
10 **ENGINEERING SYSTEMS AND INFORMATION MANAGEMENT**
  10.1 Design Software (CAD Packages)
  10.2 3D Modelling
  10.3 CAD Conform
  10.4 Progress and Performance Reporting Tool
  10.5 Piping Bulk Materials Computation
11 **COMMUNICATIONS AND REPORTING**
  11.1 Project Meetings
   11.1.1 Meetings with <Client>
   11.1.2 Internal Meetings
  11.2 Reporting
   11.2.1 Weekly Internal Reporting Requirements
   11.2.2 Client Reporting Requirements
  11.3 Communications Matrix
  11.4 Internal Project Communications Protocol
  11.5 External Communications to <Client>
  11.6 Communications with WSOs
   11.6.1 Central Engineering Office
   11.6.2 WSO # 1
   11.6.3 WSO # 2
   11.6.4 Outsourcing / Other Consultants
12 **SCHEDULE MONITORING AND REPORTING**
  12.1 Schedule Management
  12.2 Progress Measurement (Reporting)
13 **ENGINEERING CONTROLS (PLANNING AND PERFORMANCE)**

    13.1   Planning and Control
    13.2   Hold Point Management
    13.3   Dashboard Performance Reporting
    13.4   Quantities Management
14   **ENGINEERING BUDGET AND COST MANAGEMENT**
15   **BASIS OF DESIGN**
    15.1   Supporting Documents
    15.2   Key Design Criteria
    15.3   Future Expansion
    15.4   Plant Redundancy Philosophy
    15.5   Permits Approval and Owner Insurance
16   **ENGINEERING DESIGN MANAGEMENT**
    16.1   Management and Control
    16.2   Design Preparation
        16.2.1   Issued for Design
        16.2.2   Approved for Design
        16.2.3   Issued for Tender
        16.2.4   Approved for Tender
        16.2.5   Issued for Construction
        16.2.6   Approved for Construction
        16.2.7   Management of Red line Marked Up Drawings
        16.2.8   As Constructed Drawings
    16.3   Design Coordination and Interfaces
        16.3.1 Design Coordination with Contractors
        16.3.2   Vendor Design Interface
        16.3.3   Internal Interdisciplinary Design Review (IDR or Squad Check)
    16.4   Constructability Reviews
    16.5   Certification and Stamping
    16.6   Quality Control
    16.7   Quality Assurance
    16.8   Change Management
    16.9   Documentation Review and Approval by <Client>
    16.10   Design Verification and Validation
    16.11   Drawing and Document Approval Process
17   **HEALTH, SAFETY AND ENVIRONMENTAL (HSE) MANAGEMENT IN DESIGN**
    17.1   General
    17.2   Environmental Requirements
    17.3   Safety in Design

18 **RISK MANAGEMENT**
    18.1   Objective
    18.2   Risk Management Approach
        18.2.1 Design Risk Review
19 **CONSTRUCTABILITY REVIEWS**
20 **MAINTAINABILITY AND OPERABILITY REVIEWS**
21 **MODULARISATION AND PRE-ASSEMBLY**
    21.1   Construction Work packs
    21.2   Preparation of Pre-commissioning System Work Packages
    21.3   Engineering Support during Construction
22 **ENGINEERING DRAWING MANAGEMENT**
    22.1   Engineering Drawing System
    22.2   <CLIENT/CONSULTANT NAME> Drawing Requirements
    22.3   Design Approval Coordination
    22.4   Design Mark Ups
    22.5   Cancelled Drawings
    22.6   Numbers
    22.7   Red Line Marked Up Drawings
23 **ENGINEERING COORDINATION WITH OTHER FUNCTIONS**
    23.1   Coordination with Procurement
        23.1.1 Establishment of a Procurement Plan
        23.1.2   Procurement Deliverables
        23.1.3   Tender Response – Tender Evaluations
    23.2   Coordination with Fabrication Assembly
    23.3   Coordination with Project Controls, HSE and Quality
        23.3.1   Project Controls
        23.3.2   HSE
        23.3.3   Quality
    23.4   Design Office and Vendor Interface Management
        23.4.1   Design Office Interface Management with WSOs
    23.5   Design Office Interface with Client
    23.6   Vendor Data Interface Management
24 **CHANGE MANAGEMENT**
    24.1   Engineering Change Requests (ECR)
    24.2   Request for Information / Technical Queries
        24.2.1   RFI/TQs – <Consultant Name>/<CLIENT/CONSULTANT NAME> Interface
        24.2.2   RFI/TQs – <Consultant Name>/Vendors & Contractors Interface
25 **QUALITY MANAGEMENT**

- 25.1 Peer Review
- 25.2 Lessons Learnt and Continuous Improvement
- 25.3 Records Management
26 HANDOVER AND COMPLETIONS
- 26.1 As Built Drawings
- 26.2 Vendor and Design Documentations
- 26.3 Project Handover Data Books

APPENDICES (*Note: This is a Proposed list only, it is scalable and editable*)

Appendix A	List of Applicable Engineering Procedures
Appendix B	List of Applicable Standards, Specifications and Acts.
Appendix C	Engineering Deliverables Matrix
Appendix D	Engineering Review Quality Plan
Appendix E	Engineering Deliverables Per Discipline
Appendix F	Engineering Organization Structure
Appendix G	Engineering Training Plan
Appendix H	Standard Automation Applications
Appendix I	Project Stakeholders and Engineering Team Directory
Appendix J	Engineering Schedule
Appendix K	Engineering Quantity Management Flow Diagrams.
Appendix L	Drawing and Document Schedule
Appendix M	RFI/TQ Proforma

1
PROPOSED CONTENT FOR THE "**PURPOSE**" SECTION

This section provides clarity as to the "purpose" of the EMP, and need not be an elaborate section, but instead one which is concise and to the point.

Some suggested wording can be as follows:

"This Engineering Plan outlines the scope, methodology, processes, tools and strategies to be used for the execution of the engineering services to the agreed scope of works.

It further assures that the delivery of works is technically compliant, safe and complies with all legislations and requirements set out in the Contract (<contract reference number>).

This Plan also highlights the interface strategies within functional areas as well as work share methodologies which are to be adopted for the successful delivery of the project.

It is a "How" document, which provides a high-level overview and establishes the framework within which the engineering aspect of this project will be executed.

It is an integral part of the Project Execution Plan and covers the project stages from the DES (Definitive Engineering Studies) Stage to Project Execution Stage."

2
PROPOSED CONTENT FOR THE "PROJECT OUTLINE" SECTION

This section provides a high-level outline of the project, which can be found on the project proposal report.

Typically, it provides the following:

- An overview of what the project or engineering operations is all about,
- The type of facilities which are to be designed;
- The capacity and high-level specifications of these facilities;
- Major plant/equipment which are involved;
- Who the Client is, if this is a project designed for the Client;
- What is the output of the facility;
- A brief of the geographical location of the project;
- Any local considerations that are important and binding etc.

A sample write up is provided below for guidance:

<Client Name> has embarked on an expansion drive which commences with a power plant upgrade project.

This project comprises of the installation of power generation and transmission system across the <Specify location>, to support <Client Name>'s expansion of its production output from 100 Mtpa to 200 Mtpa of copper in that region.

This project, which is named as the <Specify Project Name> will include a new Combined Cycle Gas Turbine (CCGT) power plant facility, located approximately 2 km north of the town of XYZ

It will consist of two dual fuel LM6000 GE Aero-derivative gas turbines and one condensing steam turbine in a 2 x 1 CCGT configuration. Each gas turbine will produce 47 MW of electrical power, and the steam turbine will produce an additional 40 MW. The steam turbine will be powered by high and low-pressure steam generated in two once through steam generators (OTSGs) that will recover waste heat from each of the gas turbine exhausts. The steam turbine will exhaust into a forced draught Air Cooled Condenser (ACC). To smooth out the facility's power delivery across a range of ambient temperatures the gas turbine inlet air is chilled to improve overall plant performance.

In addition, the gas turbines are injected with water (SPRINT) and there will be a capability for duct firing in the OTSG."

<Client Name> has appointed XXYY as the Consultant for the project and its work scope includes site establishment works, engineering, procurement and construction management of an earthworks embankment, access road, site drainage, two GTG Packages, steam cycle equipment, including one ACC, one STG unit and all BOP, delivery of operation and maintenance training for the complete Facility, and commissioning and acceptance testing.

3
PROPOSED CONTENT FOR THE "SCOPE OF WORK" SECTION

This section needs to capture as much as detail as possible on the scope of work that is undertaken. Further it must comprise of a sub section which also lists out clearly as to any works which are excluded, titled as "Works Excluded".
This section needs to specify the "battery limits" of the scope of engineering that is undertaken.

A sample write up is provided below for guidance:

"The Scope of Design is outlined in Contract No A3599001 under the Exhibit 1, Scope of Work, Technical Requirements, Testing, Commissioning and Acceptance requirements.

This Contract includes the engineering, procurement and construction of a complete Facility, comprising of the following:

- 2 dual fuel DLE GE LM6000 PF GTG Units with inlet air chilling, unitised supplementary fired OTSGs, 1 STG, 1ACC and associated BOP, site establishment, earthworks and finishing works also detailed in Exhibit 1 of the Contract.

- Diesel fuel unloading, storage and transfer facility.

- A combined workshop, administration and control building facility (the Utilities Building", which forms part of the facility and is described in Exhibit 3 of the Contract.

Auxiliary balance of plant to support the main power island will comprise of the following plant:
- Natural gas and diesel fuel systems.
- Steam system.
- Feed water and condensate system.
- Water systems.
- Waste water systems.
- Compressed air system.
- Auxiliary boiler.
- BSDG (black start diesel generator).
- HV, LV, DC and UPS electrical primary and secondary systems.
- Instrumentation and control systems.

For further detail on the scope of design refer to Section 35.3 of the Contract. "Under the Exhibit 1, Scope of Work, Technical Requirements, Testing, Commissioning and Acceptance requirements.".

3.1 Works Excluded

(Specify an outline of the exclusions – a sample write up is shown below)

"The following scope is excluded:

- Fuel for the operation of the facility during commissioning, testing and operation.
- Supply of water to the facility during construction, commissioning, testing and operation of the new facility.
- Main generator step up transformer and HV switchyard.
- Gas delivery station and lateral piping design, supply and installation.
- Grid connected load providing power demand for commissioning, performance and reliability testing.

- *All employer supplied equipment described in <specify section referred in Contract e.g. Section 5.0 of Exhibit 1 of the Contract>.*
- *<Client> to arrange the following approvals.*
- *Heritage Approval;*
- *Biological Approvals (Flora and Fauna);*
- *Tenure (define the site boundaries);*
- *State Development Agreement;*
- *Water Licenses for construction;*
- *Clearing Permits;*
- *Part V Approvals of the EPA Act Stage 1 (works approvals and bed and banks for construction)*
- *Part V Approvals of the EPA Act Stage 3 (Commissioning compliance for granting operating license);*

4
PROPOSED CONTENT FOR THE "TERMS AND DEFINITION" SECTION

This Section should contain definition and clarification of all the acronyms that will used throughout the EMP.

It is best that these are tabulated for ease of reference, a sample of this is as shown below:

Term	Definition
Procedure	Specified way to carry out an activity or a process.
Specification	Document stating requirements.
Contractor/Supplier	Party who has been awarded a contract or

Term	Definition
	purchase order by <Client Name> for the Project.
IDR (Inter-Discipline Review)	Internal "squad check" of drawings/documents by multi-discipline engineering individuals or groups.
CAD	Computer Aided Design
Client Review	Review by Client after Inter-Discipline Review.
DES	Definitive Engineering Study
EIP	Estimate Information Pack
EPCM	Engineering, Procurement and Construction Management.
ITP	Inspection and Test Plan.
MDR	Manufacturer's Data Report -a document or collection of documents, providing objective evidence that specified requirements have been satisfied, to the extent required by an approved Supplier or Purchase Order. Documents include but are not limited to, inspection reports, test reports, non-destructive examination reports, and certificates of compliance, material test reports, and the like.
PCM	Procurement and Contracting Manager.
WSO	Work Share Offices
WBS	Work Breakdown Structure etc. etc.

5
PROPOSED CONTENT FOR THE "REFERENCES" SECTION

This section tabulates the references that the EMP makes to other pertinent and crucial documents throughout the pan.

A sample reference table can be as follows:

Document Number/Title	Description
XX-YYY-ZZ	List of Contract Deliverables Specifications
XX-YY-ZZ	Project Execution Plan
XX-YY-ZZ	(Related) Document Control Procedures
XXX-YY-ZZ	(Related) Project Control Procedures
Appendix A	List of Engineering Procedures
Appendix B	List of Applicable Standards, Specifications and Acts
etc.	etc.

6
PROPOSED CONTENT FOR THE "APPLICABLE PROCEDURES, STANDARDS, SPECIFICATIONS & ACTS" SECTION

This section specifically highlights all the applicable procedures, standards, specifications and acts which are applied to the engineering process of the project/operations.

A sample write-up can be as follows:

6.1 Engineering Procedures

The engineering services shall be executed in accordance with the following procedures:
- relevant <Client Name> procedures;
- <Consultant Name> engineering procedures;
- as well as any project-specific engineering Instructions/procedures and/or guidelines which may be developed as required.

These are listed in **Appendix A**

6.2 Standards and Specifications

The applicable standards and specifications shall be referenced to in delivering the engineering scope of works in this project.

These are related to the following:
- <Country Specific> Standards;
- Applicable International Standards;
- <Regional> Government Acts;
- <Client/Consultant> Standards and Specifications.

These are listed in **Appendix B**

7
PROPOSED CONTENT FOR THE "SCOPE OF ENGINEERING SERVICES" SECTION

The section outlines as detailed as possible the scope of the engineering services which are undertaken on this project.

A sample write up for this section can be as follows. Notice that this sample write up gets right down to the three (3) sub sections, however it can be more depending on the scope and complexity of the projects/operations.

7.1 Disciplines Involved

This scope which comprises of design and drafting will be inclusive but not limited to the following disciplines:
- Mechanical;
- Process;
- Civil;
- Structural;
- Architectural;
- Electrical;
- Control Systems and Instrumentation (including Communication and Programming); and
- Hydraulics/Pneumatic.

7.2 Engineering Deliverables

All engineering deliverables will be generated, first to procure the necessary equipment and bulk commodities and secondly to execute construction contracts.

The scope for individual construction contracts will be defined so that engineering deliverables will be produced with minimum conflicts across different contracting packages in accordance with the strategy outlined in the project's contracting and procurement plan <specify plan doc number>.

This plan also contains a complete list of contracts which are to be awarded on the project.

The project document control team will capture, store and transmit engineering documents and related correspondence in accordance with the project document management procedure <specify procedure doc number>.

The engineering team will produce and maintain the following documents that encompass the deliverables package to be produced by the Engineering Team:

- Engineering Deliverables List
- Engineering Schedule (incorporated as part of the main schedule)
- Engineering Management Plan
- Procurement List
- Vendor Data / Document Register

These documents will be developed and maintained by Engineering Leads to ensure engineering is planned, monitored, measured and controlled throughout the lifecycle of the project.

This ensures successful implementation of engineering to the agreed scope of works as well as delivery of works that is technically compliant, safe and complies with all legislations and contractual requirements.

This deliverables list is the basis for the Level 4 engineering schedule and engineering progress measurement.
Each discipline lead will review and update the deliverables on an ongoing basis and agree changes with the Engineering Manager.

Where appropriate, engineering man hour budgets will be adjusted to reflect the revised deliverables and will be issued in the form of a change management process.

Refer to **Appendix C** for the Engineering Deliverables Matrix.

7.3 Engineering List of Work Activities

Consistent with the scope of the Contract, the list of work activities and deliverables include:

- Producing all discipline engineering deliverables and tasks, such as design criteria, calculations, design drawings, technical and general specifications, material requisitions, services requisitions, equipment lists, electrical load lists and line designation and valve lists.

- Developing quantities of all engineered and bulk commodities.

- Coordinate the design with other business units, work share offices, work generating offices, consultants and others to ensure facility quality requirements are met with respect to operation and maintenance.

- Develop the technical portion of contract packages, including assembling pertinent drawings, data sheets and construction specifications, and support contracts and procurement in preparing the appropriate scope definitions, pay items, and schedules.

- Develop Process Flow Diagrams (PDFs) and Process & Instrument Diagrams (P&IDs).

- Develop design criteria and standards for each discipline.

- Implementation of the modularization initiative < refer to any modularization guideline if available>.

- Develop material requisitions, service requisitions, specifications, and data sheets for purchase of the plant equipment.

- Perform technical bid evaluations of proposals for equipment and material purchases and recommendations for purchase.
- Develop material requisitions and specifications for purchase of bulk materials based on material quantity take-off for the designs.

- Review vendor drawings and documents for compliance with scope and specifications, as well as identify all design interface requirements for preparation of project deliverables.

- Populate the progress and performance report system consistent with the detail engineering schedule and the project scope for drawings, specifications, material requisitions, service requisitions and engineering tasks.

- Assist with the development of any cost estimates from which project performance can be monitored.

- Assist with the functional quality reviews to be performed as part of the Engineering Review Quality Plan – refer **Appendix D**.

- Design quantities and manage design information engineering quantity changes and scopes changes as per the Change Management Procedure <specify document number>.

- Use an Engineering task list to manage all tasks and provide detailed reports for use by end users <specify the software system adopted if one is utilized>

- Conducting optimization and value improvement studies such as constructability, operability, maintainability, energy optimization of waste minimization and others.

- Provide input to the Project Execution Plan, Project Controls Plan, Project Quality Plan, Project Construction Plan etc.

- Additionally, under each discipline, resources will be allocated for "tasks" to manage various issues such as:

    - design optimization;
    - design reviews;
    - HAZID, HAZOP, RAMBO, SIL Studies etc.;

- desk reviews (these will be conducted with the vendors), special vendor print review and coordination; and
- inter-discipline coordination, calculations, administration, meetings and special studies.

7.1.1. Approximate Deliverables for Each Discipline

Deliverables will be generated for each of the following disciplines:

- *Process Engineering;*
- *Mechanical Engineering;*
- *Mechanical Design and Drafting;*
- *Piping;*
- *Piping Design and Drafting;*
- *Electrical;*
- *Control Systems;*
- *Civil/Structural/Architectural.*

The approximate number of deliverables for each of the disciplines is shown in **Appendix E.**

8
PROPOSED CONTENT FOR THE "ORGANISATION" SECTION

sample write up for this section can be as follows:

8.1 General

The engineering design work will be allocated based on the following criteria to bring the best value to the Project in terms of:

- *Local legislative requirements related to registered local engineers and architects;*
- *Resource availability and proven expertise;*
- *Geographical location(s) of the project site(s);*
- *Schedule considerations*
- *Risk management;*

- Client proximity and requirements;

The team shall be structured by discipline as per the following:

- Engineering Coordination and Planning
- Design & drafting;
- Civil, Structural and Architectural;
- Mechanical;
- Electrical, Controls and Instrumentation;
- Process;
- Hydraulics etc.

The Engineering Manager in consultation with the Project Manager shall be responsible for reviewing and approving all design engineering resources for the projects.

It is to be noted that the review and approval of project engineering requirements shall be the responsibility of the respective Area Managers in consultation with the Project Manager.

The Engineering Manager shall be solely responsible for the design and technical compliance of the project. He works closely with the Area Managers.

8.2 Locations

The primary engineering team will be based in <specify location> and will provide engineering management oversight to all parties including Work Share Offices (WSO), external consultants, vendors and contractors performing engineering work.

During the execution phase of the project, a core engineering team shall be established on site to coordinate and facilitate engineering design requirements with the site construction team and contractors.

8.2.1 Work Sharing Plan

A work sharing plan shall be developed, which details the responsibilities of each engineering area and discipline, and an outline on who is responsible for producing relevant deliverables.

<Provide a link to the draft Work sharing matrix here if available>:

Deliverables by other offices will be provided in the agreed communications method to the Engineering Manager as per the agreed format, stages and deliverable packages, with all copies of the design documentation stored within the respective project folders within the central engineering office.

The deliverables list for engineering will be the responsibility of the Engineering Manager to ensure that they are met and are in accordance with the Contract requirements with respect to scope, quality, time and budget.

The communications and information management process are elaborated further in the relevant section of this Plan under "Communications & Reporting".

8.3 Engineering Organization Structure

The organization structure of the engineering team can be referenced in **Appendix F**.

It describes the reporting relationship within engineering team as well as delineation of responsibilities, which includes the WSOs.

The Engineering manager reports to the Project Manager, and together with the lead engineers, they are responsible for the project design deliverables, meeting the baseline milestones regarding budget, schedule, and deliverables.

The primary roles identified to conduct the engineering design for this project, and principal functions are described in detail in the section below.

8.3.1 Roles and Responsibilities

The following are the key roles within the engineering design team:

- Engineering Manager;
- Principal Discipline Engineers;
- Lead Discipline Engineers;
- Senior Designers; and
- Drafting Office Manager.
-

8.3.2 Engineering Manager

The Engineering manager is accountable for the delivery of the Engineering component of the project in accordance with the agreed Project milestones, schedule and design plan.

Specific responsibilities:

- Develops the project engineering scope document and deliverable plan for engineering;
- Develops and monitor the project engineering budget;
- Assists with the development of the project design plan;
- Manages the design program and ensure deliverables are available on time in accordance with the project schedule;

- Outlines the design team structure needed for the project based on the engineering scope budget, schedule and design plan;
- Develops the project procurement strategy in conjunction with the Engineering Discipline;
- Leads, plans, facilitates and closes out HAZID/HAZOPs as required;
- Reports project design activities and deliverables progress;
- Develops the detailed engineering schedule in conformance with the project schedule and milestones;
- Responsible for overall plant design compliance;
- Ensures project resourcing meets the overall project demands;
- Appoints and lead the tender response and evaluation process teams;
- Plans and manage the design risk identification and mitigation process for the project;
- Plans, co-ordinates and reviews project design activities and deliverables;
- Technical management of design sub-consultants;
- Coordinates approval of design documentation for construction and procurement;
- Resolves interdisciplinary and interface issues;
- Co-ordinates internal document and drawing squad checks and reviews;
- Progress reporting and liaison with the Project Manager; and
- Provide forecasts of resourcing needs for delivery of project outputs.

8.3.3 Principal Discipline Engineers

The Principal Design Engineers report to the Engineering Manager and are primarily responsible for the review and approval of deliverables as well as providing the direction to their respective teams related to the following:

- Responsible for allocated discipline design packages and associated deliverables.
- Approval of discipline TQs and RFIs.
- Ensure discipline resourcing meets the overall project demands.
- Manage interface and coordination of the discipline engineering team.
- Technical management of design sub-consultants.
- Technical approval of discipline design documentation for construction and procurement.
- Plan, co-ordinate and review discipline design activities and deliverables.
- Ensure compliance with critical project requirements and engineering plan.
- Progress reporting and liaison with the Engineering Manager.
- Participates in design reviews and facilitates internal discipline reviews.

8.3.4 Lead Discipline Engineers

The Lead Discipline Engineers report to the Principal Discipline Engineers and they shall be responsible for the following:

- Support the Principal Discipline Engineer by being the point of contact for the respective engineering discipline on this project
- Report progress weekly to the Principal Discipline Engineer and Engineering Manager.
- Conduct mechanical design work, engineering reviews and assistance throughout construction phase on site.
- Manage and conduct activities such as planning, manpower resources/tools assignment and technical assistance across projects.
- Forms part of the Mechanical design team and is only required for the duration of the project/s.
- Attend all design and management meetings as required.
- Carry out the electrical design activities in an efficient and professional manner.
- Review designs, specifications and procedures as required.
- Ensure all controlled documentation/drawing is current.
- Take part in audits, design reviews and design verifications as required.

8.3.5 Senior Designers
Coordinate the design review process across the design offices towards the following:

- Produce detailed design and drafting of mechanical elements within the scope of work;
- Develop all 2D and 3D modelling designs; and
- Mentor and supervise designers under their respective disciplines.

8.3.6 Drawing Office Manager
- Develop tools for productivity and ensure that all drawings are presented in a uniform manner.
- Install and support all CAD related software for production and drawing viewing.
- Ensure completed latest revision drawings are delivered to the Client.
- Liaise with share offices for cad support.
- Liaise with vendors to ensure they meet the required level of Client cad compliance.
- Conduct strategic planning for software versions, implementation and installation.
- Work with Bentley Systems on product direction and evaluation.
- Coordinate and liaise effectively with engineers and managers.
- Liaise with Drafting Agencies to hire drafting resources.
- Create and maintain technically consistent drafting methods.
- Adapt to continuously changing CAD packages.
- Provide time and resources for necessary in-house or external training.

8.3.7 Training and Development

As far as practically possible, only competent and experienced personnel are sourced and recruited into the engineering team who can contribute meaningfully towards the project within the shortest possible time frame upon undergoing the project internal induction program.

To further develop and enhance the technical and overall design delivery competencies of the team, there are various modes of training that would be made available within this project.

These range from:

- On the job training through mentorship and close supervision, which is the primary mode;
- Specialized technical training related to any new software systems or design development tools which the team may not familiar with;
- Procedural (both of <Consultant Name> & Client) awareness program; and
- Design delivery process and risk review process.

The Engineering Manager, Principal and Lead Engineers shall supervise and mentor all newly appointed team members.

They shall further ensure that the required project inductions as directed by the Project Human Resources Section have been effectively undertaken by every member of their team.

8.3.7.1 Engineering Training Plan

The Engineering Manager in consultation with the Principal and Lead Engineers may develop a training plan encompassing the training requirements and framework for the project engineering team.

This plan will be based on identified training needs (both technical and soft skills) of its resources. The training plan, to be maintained current, will list all people and suggested training and a schedule for everyone. The work share offices shall develop a similar training plan for their resources.

Refer to **Appendix G** for the Engineering Training Plan.

9
PROPOSED CONTENT FOR THE "WORK BREAKDOWN STRUCTURE (WBS)" SECTION

The WBS is the heart of the overall monitoring and tracking of the progress with the engineering design process.

Hence, due care and diligence must be accorded in the development of the WBS and assignment of the WBS codes.

A sample content of this section can be as follows:

The project WBS is in the form of a project code of accounts, project commodity listing and a Work Packaging Plan which will provide the framework to identify and categorize quantities, man hours and costs related to project management, engineering services, construction materials and construction installation.

These are established in discussion and agreement with the client.

They are updated on a regular basis to reflect its latest status at the following folder location<specify folder location or paste live-link to this folder here>.

This WBS structure has been established in line with the <Project Name> Standard XX-YY-ZZ (specify the doc reference number here, if available) as follows:

- *Project (prefix) code (<Client> Projects Level);*
- *Project phase (<Client> Projects Level);*
- *Project location (<Client> Projects Level);*
- *Facility (CMS Level); and*
- *Commodity (CMS Level).*

The breakdown of the facility and commodity codes used on the <specify project name> is summarised as follows:

	TABLE 1 – WORK BREAKDOWN STRUCTURE	
Level Name	Description	Example
L1 Project	Level 1 represents the Project.	4000 – Port
L2 Area	Level 2 represents the functional area of the Project.	41 In-loading 42 Stockyard/Stacking 42 Reclaiming Etc
L3 Facility	Level 3 identifies particular facilities with an Area.	41-21 Conveyor C450
L3a Sub Facility	Level 3a represents the 5th character WBS element which is used to identify	48-21 Water 48-21(1) Process Water

	any sub-facilities required for further breakdown of the estimate, dependent on the magnitude of the Project.	48-21(2) Potable Water 48-21(3) Recycled Water
L4 Engineering Discipline/Commodity level 1	Level 4 represents the type of commodities used in the capital cost estimate/resource disciplines used during the design phase of the Project. It is also the level at which construction actual hours are collected.	
L5 Planned Task/Activity	Level 5 of the WBS represents the planned engineering or construction activities in the Project master schedule and is also the level at which engineering actual hours are captured.	
L6 Deliverable / Commodity pay line items	Level 6 of the WBS is the level at which budget is held and is also the level at which progress is captured.	

9.1 Deviations

<Specify clearly if there are any deviations that have been adopted from the approved WBS at any stage of the project, highlighting clearly justifications for doing so.>

10
PROPOSED CONTENT FOR THE "ENGINEERING SYSTEMS AND INFORMATION MANAGEMENT" SECTION

This section details out the automation applications that are used for the design and data management purposes.

A sample write up for this section can be as follows:

"The Engineering team will utilize the following standard automation applications as listed in Appendix H for its design and data management purposes.

Waiver from use of specified applications requires prior approval from the Engineering Manager and/or Project Manager.

10.1 Design Software (CAD Packages)

CAD Design Software is used by Engineering with specific design modules when required.

The Contract requires the drawings delivered in CAD format however drawings will be produced during the performance of the design work in MicroStation or AutoCAD. A final conversion to CAD will be performed prior to hand over to <Client>.

10.2 - 3 D Modelling

The 3D SmartPlant system will be used as the design tool to generate many of the project deliverables required for specific areas of the plant.

The Project drawings will be produced primarily using < specify software >. All drawings will be in < specify software > format.

A 3D model of the process facilities will be completed using < specify software >.

This model will first be established to concrete outlines, structural steel framing elements and major mechanical and electrical equipment, and plate work.

Major electrical raceway and most piping will be designed in a 3D environment driven by established databases and included on the 3D model.

The 3D model will be used primarily for interference checking (steel, concrete, raceway, piping, and plate work) and preparation of structural steel drawings and piping orthographic and isometric drawings.

The 3D Model Design Review software automation will be used for on-Project, HSEC Reviews (Safety by Design) and design reviews.

The project area lead designer will be responsible for management, control, coordination, and assurance of the integrity of the model.

10.3 CAD Conform

CAD Conform is an external software module that can be loaded onto either AutoCAD or MicroStation drafting package.

It is used to check that all CAD format drawings submitted to the <Client> for approval conform to the client's drawing standards i.e. all drawings submitted to the <Client> representative by Engineering must conform to the <Client> Drafting Standards.

10.4 Progress and Performance Reporting Tool

Engineering planning and control will be done via a Progress and Performance Reporting Tool <specify the name of this specific tool that is used if available>.

This tool will be developed and maintained in accordance with project control requirements.
This tool will be used to track the progress of engineering deliverables and tasks and will highlight variances between actual hours required and budgeted hours.

In monitoring and managing engineering, a detailed schedule will be established and maintained up to approximately 90% engineering completion.

This tool will further be coded to monitor the above scheduled activities and will be up-dated weekly as per the schedule and fortnightly for the progress and performance reporting purposes.

10.5 Piping Bulk Materials Computation

Piping bulk materials for the Project will be tracked and monitored by use of computerized material control program <please specify>."

11
PROPOSED CONTENT FOR THE "COMMUNICATIONS AND REPORTING" SECTION

This section outlines the whole communications and reporting framework that is established for the effective execution of this project/operations.

A sample write up can be as follows:

"The successful implementation and continuous improvement of the Project relies largely on effective communication lines being developed and maintained at all levels of the project structure.

This section serves to set out the engineering communications and reporting framework for this project. It will serve as a guide throughout the life of the project and will be updated if and when project needs change.

It includes a communications matrix which maps the communication requirements of this project.

A project team directory is included in Appendix I to provide contact information for all stakeholders and key engineering team members who are directly involved in the project.

11.1 Project Meetings

11.1.1 Meetings with <Client Name>

Fortnightly Engineering Interface meetings shall be held between <Consultant> and <Client Name>.

These meetings shall review and discuss issues related primarily to the following:
- Engineering design progress;
- Deliverables interface constraints;
- Critical path issues;
- Safety in Design issues;
- Resources
- Risks etc.;

All meetings and meeting records between <Client Name> and <Consultant> will be coordinated and documents maintained within the project database.

These records shall be submitted to the relevant parties within <Client Name> and <Consultant Name> for approval prior to distribution to the meeting participants through the <Consultant> Correspondence Logging database.

11.1.2 Internal Meetings

The following are the suite of internal meetings which the engineering team is involved with within the project:

- Project Team Meetings (weekly).

- Engineering Team meetings (weekly).
- Project Schedule Meetings (weekly).
- Commercial Meetings (weekly).
- Project Risk Identification and Management meetings (monthly).
- Construction onsite meetings (Tool box) daily (during construction phase).
- Other meetings, as and when required.

All meeting shall be coordinated in advance with an agenda provided to all attendees outlining the purpose and objectives of the meeting.

All meetings shall be formally recorded with an action log prepared and maintained by the Project Admin.

Templates for meeting records and action logs can be located on <Consultant/Client> database.

For internal meetings, copies of meeting records shall be distributed to all attendees.
All meetings records shall be kept and maintained within <Consultant Name> database for traceability and auditability purposes.

11.2 Reporting

11.2.1 Weekly Internal Reporting Requirements

The Engineering Manager shall provide to the Project Manager a weekly engineering report on engineering progress highlighting:

- Engineering progress – planned and actual;
- Progress on deliverables;
- Budget;
- Scope changes;
- Issues; and
- Risks.

11.2.2 Client Reporting Requirements

These reporting requirements comprise of weekly and monthly reporting, which are described below.

11.2.2.1 Weekly Reporting

The weekly progress reporting would typically be the similar to the weekly internal reports.

These reports are coordinated through the Engineering Manager to the Project Manager before they are submitted to the Client.

11.2.2.2 Monthly Reporting

The monthly progress reports provide a more comprehensive overview over and above the weekly reporting in relation to the following:

- HAZID/HAZOP review progress;
- Design Review progress;
- Interface Coordination and Management;
- Management of Change;
- Resource utilization review; and
- Etc.

11.3 Communications Matrix

The following table highlights the communications matrix for the project in general. The engineering reporting and communications requirement shall be governed in accordance with this matrix.

Target Groups	Communications	Frequency	Methods of Communication	Responsibility
<Client or Consultant Name> Project Team	Project progress reports; &	As requested.	Project Status Reports. Report Presentations Emails/Minutes of Meetings	Project Manager

	Key messages and updates to Project Team and Client Team on project status	Fortnightly/ monthly	Emails/Minutes of Meetings	Project Manager
<CLIENT Mgt Team>	Weekly Project progress report & Monthly Project progress report	Weekly Monthly	<ul><li>Project Status Reports</li><li>Meeting minutes</li><li>Formal emails via Coro Logs.</li></ul>	Project Manager
	Issues and reporting highlights for resolution.	As required.	<ul><li>Reporting and Rectification Reports</li><li>Aconex</li><li>Formal/Informal emails/verbal comms</li></ul>	Project Manager and/ or Project Director
Consultant Mgt Team	Progress Reports and updates.	As required.	<ul><li>Project Status Reports.</li><li>Report</li><li>Presentations</li><li>Meetings</li><li>Minutes</li></ul>	Project Manager and/ or Project Director

Engineering	Engineering Progress Reviews Design Reviews/ HAZOPS	Regularly during design phase	• Scheduled design meetings • Transmittals • Workshops • Meeting minutes • Project status • Reports • Drawings and Documents • Deliverable Plan	Project Manager Engineering Manager
	Design Progress	Regularly during design phase	• Scheduled Mtgs • transmittals • Workshops • Meeting minutes • Project Status reports • Payment Certs.	Project Manager. Engineering Manager
Project Mgt Team.	Issues requiring resolution.	As required.	• Directives. • Weekly Meeting with Commercials • Meeting with Project Sponsor • Project Reporting (highlights)	Project Manager

11.4 Internal Project Communications Protocol

All project personnel shall use the correspondence database <specify name> to capture and/or transmit project communications such as project documentation, deliverables, decisions and issues pertaining to the project.

With emails, the formal project email address <specify> shall be used to ensure capture of all email strings pertaining to the project.

The engineering team will coordinate their design revision and review process using <specify system> however once designs have been approved these will be issued via Project Document Control to the client utilizing the <specify the document control system>.

11.5 External Communications to <Client>

External communications to the Client will be coordinated through the through the project correspondence database <specify name> for emails and uncontrolled documentations, whereas all controlled documentations e.g. drawings, specifications, schedules etc. shall be issued via Project Document Control utilizing the <specify the document control system>.

The process of requesting distribution of documents via Project Document Control shall be governed by the procedure <specify>.

11.6 Communications with WSOs

The flow chart below describes the method of how documents will flow between the design areas and design offices in a coordinated approach (Insert the flowchart relevant to the project).

The sub sections below provide a brief overview of the key responsibilities of each WSO in supporting the central engineering office:

11.6.1 Central Engineering Office

The central engineering office will be responsible for the overall control of the engineering effort and all documents will be authorized and issued through document control based in the <specify location> office.

11.6.2 WSO # 1

The WSO located in Chile will undertake the following scope of engineering design services:

- (specify these – and attach a high-level deliverables list where possible);
- (highlight as to how engineering QA/QC will be managed especially in relation to design reviews/ H
- Hazids / SiDs etc)

11.6.3   ***WSO # 2*** *(etc. etc.)*

11.6.4   ***Outsourcing / Other Consultants***

Several specialist consultants will be used to supplement engineering expertise and verify the detailed design.

The key consultants to be engaged are regarding the following design packages:
- Boiler Chemistry
- Boiler and Steam System Review
- Earthing Design
- Lighting Design
- Utilities Building Services
- Utilities Building Structural Design

These consultants are either pre-qualified or have credible reputations in the market for delivering the required services.

Their nomination and selection shall be reviewed with the Client before they are formally engaged.

They shall be engaged in accordance with <Client Name> Corporate sub consultancy agreements ref <specify if available> and shall conform to the requirements of the Contract for this project."

12
PROPOSED CONTENT FOR THE "SCHEDULE MONITORING AND REPORTING" SECTION

A proposed sample write-up for this section can be as follows:

12.1 Schedule Management

Engineering Schedule and milestones are developed by Engineering Manager/ Principle Engineers and Lead Discipline Engineers and agreed at the start of the project.

Engineering will develop key milestones in conjunction with contractual milestones, which will be incorporated into the main project schedule.

Engineering milestones are tracked and monitored on a weekly basis as part of regular progress report by the Engineering Manager.

For accurate copy of the project schedule and milestones please refer to **Appendix J**.

12. 2 Progress Measurement (Reporting)

Progress measurement will be required for reporting purposes and a percentage complete has been allotted for the various stages of document completion.

At the end of each reporting period the progress for each document group is reviewed by the Engineering Manager to verify progress and to initiate any recovery program, as required, for any areas of concern.

The Engineering deliverables schedule for the project will be used to measure progress of Engineering.

Actual engineering document progress information will be drawn from the document control system and used to update Engineering progress.

The actual percentage completion of each document / design activity within a work package is assessed on milestone scales as noted.

The standard milestone gates for engineering deliverables together with their physical percent completion are listed below.

The Lead Discipline Engineers in consultation with the Principal Engineers will provide updates to Project Control either directly or via the Engineering Manager with forecast dates to completion.

The Project Planner will confirm that the combined result of the remaining durations, forecast dates to complete and engineering logic.

Where there are conflicts, these will be resolved with the Engineering Manager, Principal Engineers and Lead Discipline Engineers.

Proposed Milestone Gates for Engineering Deliverables Table

Discipline	Document Type	Start	Rev A	Rev B	Rev C	Rev 0
Civil	CALCS	15%	60%			100%
	DRWG	15	60	80		100%
	REPORT	15	60	80		100%
	SPECS	15	60	80		100%
Electrical (similarly, for Control Systems Mechanical, Piping, Structural, Architectural etc.)	CALCS	15	60			100%
	DRG	15	60	80		100%
	ITP	15	60			100%
	MDR	15	60			100%
	REP	15	60	80		100%
	SCE	15	60	80		100%
	SCH	15	60	80		100%
	SLD	15	60	80		100%
	SPECS	15	60	80		100%
	TD	15	60	80		100%

13
PROPOSED CONTENT FOR THE "ENGINEERING CONTROLS (PLANNING AND PERFORMANCE)" SECTION

A proposed sample write-up for this section can be as follows:

13.1 Planning and Control

"Engineering planning and control will be done via a Progress and Performance Reporting Tool <specify the name of the specific tool that is used>.

This tool will be used to track the progress of engineering deliverables and tasks and will highlight variances between actual hours required and budgeted hours.

In monitoring and managing engineering, a detailed schedule will be established and maintained up to approximately 90% engineering completion.

This tool will further be coded to monitor the above scheduled activities and will be up-dated weekly as per the schedule and fortnightly for the progress and performance reporting purposes.
The deliverables and tasks that define the engineering scope will be listed in the Progress and Performance Reporting Tool logs and rolled-up per activity type in the detailed engineering schedule.

Every engineering deliverable and task will have an assigned office, a job-hour budget, a responsible engineer/ designer, a PO and contract number, and schedule dates for review, approval, and issue for construction.

The progress reports will be extracted from the Progress and Performance Reporting database to determine engineering progress in terms of scheduled delivery of design documents and task performance against budgeted engineering job-hours.

Look-ahead reports will be used for short term planning.

Project controls will assign a Planner to assist the engineering group in the above task.

Only one Progress and Performance Reporting database will be established for the Project, listing deliverables and tasks of all offices.

External consultants will provide their input to the Progress and Performance Reporting Tool as well as the WSO, who will have access to the Progress and Performance Reporting Tool and will be responsible for entry of progress and dates in collaboration with the Planner.

13.2 Hold Point Management

A hold point tracking tool shall also be used to monitor finalisation of deliverables. The hold point management is part of the engineering quality control initiatives.

13.3 Dashboard Performance Reporting

Engineering performance will be measured against budget in four key areas, which are schedule progress, bulk quantity costs, engineering services cost and job-hour performance.

To arrive at an overall performance factor, each of the four parameters will be weighted according to its importance to overall objectives.

The indicative weighting values that will be adopted are as follows:

Schedule performance – 40%.
Total quantity performance – 40%.
Engineering cost performance – 10%.
Job-hour performance – 10%.

The WSOs will provide their input on the above parameters to the main office to produce a consolidated monthly engineering performance report in the form of a "dashboard".

13.4 Quantities Management

Engineering has the primary responsibility to provide the specification and quantities of bulk materials.

Piping bulk materials for the Project will be tracked and monitored by use of computerized material control program <e- Q tracker or equivalent>.

For facilities designed in the 3D environment, isometrics will be generated, and a direct computer link will be established for the automatic transfer and updating of piping material quantities.

For facilities designed in the 2D CAD environment, isometrics, and material lists will be prepared for all 65 mm and larger carbon steel pipe. Fifty (50) mm and smaller pipe will be field run from P&ID drawings.

Plate-work material requirements will be listed on drawings material quantities and summarized for procurement.

Electrical bulk materials will also be tracked and monitored using a material take-off program <e- Q tracker or equivalent>.

The summary information will be by area, facility, contract, construction work pack etc. as appropriate.

Instrument installation detail and material lists will also be summarized by area, facility, contract, and construction work pack and commissioning system.

*The processes adopted in relation to the management of quantities is shown in **Appendix K – Engineering Quantity Management Flow Diagrams.**"*

14
PROPOSED CONTENT FOR THE "ENGINEERING BUDGET AND COST ESTIMATION" SECTION

A proposed sample write-up for this section can be as follows:

The Engineering Manager is responsible for managing and controlling the engineering budget within this project.

The Engineering Manager together with the Principal and Lead Discipline Engineers will provide the project cost controller with the list of deliverables and discipline specific man hours to enable the project to forecast its costs with regular monthly updates.

The man hours for engineering on this project are estimated to be <specify approximate estimate> man hrs. (NOTE: It is best to provide a live link here to the current estimated remaining man-hrs so that the information on this document remains current)
The Engineering Manager is accountable for monitoring and tracking deliverables/man-hours against planned forecasts figures via regular weekly and monthly reports to the project manager.

Forecast cost and schedule to complete will also be projected and provided to the project cost controller for overall tracking and reporting purposes."

15
PROPOSED CONTENT FOR THE "BASIS OF DESIGN" SECTION

A sample write up of this section can be as follows:

The design input shall be as per the requirements of the Contract and relevant <Country> Standards and Specifications as well as <Client/Consultant Name> generated discipline specific basis of design requirements.

15.1 Supporting Documents

At minimum, the following shall comprise as the supporting documents for the basis of design:

- Process calculations sizing basis;

- Mechanical and piping specifications and calculations;
- Civil/Structural basis of design;
- Electrical basis of design;
- Instrumentation basis of design;
- Vendor document and data submittal schedule (VDDSS);
- Project Engineering Plan (PEP) this document; and
- Project schedule.

15.2 Key Design Criteria

The key design criteria for this project are as tabulated here below.

These are determined about the requirements of the Contract specifically about sections <specify relevant sections of the contract>.

(The table below provides a sample key design criteria)

Design Element	Design Criteria
Facility net output at 30°C and 40%RH	130MW
Facility net output at 45°C and 20%RH	110MW
Civil and structural works design life:	50 years
Operating design life:	25 years
Environment:	cyclone area, coastal and dust
Performance design ambient condition:	30°C and 40%RH
Maximum operating design ambient condition:	50°C and 10%RH
Minimum operating design ambient condition:	10°C and 90% RH
Approximate site elevation	10m
Seismic design class:	III
BCA importance code:	4
Design wind speed:	100m/s
Operating regime:	Base load
Number of starts per day:	one
OTSG dry operation maximum cumulative time	16250 hours
Near field noise limit – sound pressure level	85dBA at 1m distance and 1.5m height
Far field noise limit – sound pressure level	60dBA at 125m
Natural gas system minimum design pressure	6.2MPa(g)
Natural gas system maximum demand pressure	5000kPa(g)
Demineralized water storage capacity	8 hours
Diesel fuel bulk storage for GT full load operation	6 days

Potable water storage capacity	24 hours
Raw water storage capacity	24 hours
Fire water storage capacity	24 hours
Evaporation pond containment of process water and 10-year ARI event	6 hours
Evaporation pond max permeability	2.8×10^{-3} m/day
Oil spillage bund sizing	110% plus fire water allowance
Emergency DC power supplies	8 hours
UPS support for C&I system	4 hours

(Further sample write up is as follows to give more guidance)

Additional design parameters include:

The plant overall drainage system must have protection from flows of 1 in a 100-year recurrence interval rainfall intensity event discharge from the Site without rising within 300 mm of the lowest building floor level. Drainage off and around the site pad must be incorporated in the drainage design. Note specific freeboard requirements of up to 500mm for plant and equipment. The local plant drainage system, unless noted otherwise, must be designed for five-year recurrence interval rainfall intensity for minor catchments and 20-year recurrence interval rainfall intensity for major catchments.

The evaporation ponds must be designed to contain the process waste water flows from the Facility, and rain water arising from at least a 10-year ARI event for 6 hours.

Stairways and platforms need to be provided to all elevated equipment that requires access for O&M. Ladders only to be provided as a secondary means of emergency egress.

Geometric design of roadways must comply with the current <Country Name> roads standards. Road geometry must be designed in accordance with <Country Name> – Guide to Geometric Design of Rural Roads.

Road Intersections must be designed in accordance with <Country Name> – Guide to Traffic Engineering Practice. Part 5, Intersections at Grade.

Pavements must be designed in accordance with <Country Name> – Pavement Design Manual or MRWA Procedure for the Design of Flexible Pavements – Engineering Road Note 9.

Suitable stabilised and screened bedding and backfill materials, subject to the approval of the Employer, must be provided for all direct buried services. Stabilised backfill of.

<1.2°Cm/W, in accordance with AS/NZS 3008.1.1, must be provided.

Electrical buildings must be equipped with VESDA fire detection systems and inert gas flooding fire suppression system, with FM200 as the inert gas employed.
Insulation on surfaces must result in touch temperatures of less than 60°C (for outdoors based on 25°C ambient in the shade). In areas not requiring insulation for heat conservation, where the surface is accessible to personnel, is within 2 metres of any floor or access way and where insulation is impractical, then, as an alternative to insulation, stand-off protection may be provided.

Building/s must be provided to house the following:

- ✓ Water Treatment Plant
- ✓ Compressed air equipment
- ✓ Dangerous goods storage

15.3 Future Expansion

(Highlight and specify in this section any future expansion requirements which is involved in the scope of work)

15.4 Plant Redundancy Philosophy

(Sample write up is as follows)

The overall plant redundancy philosophy is as follows:

Primary unitised Plant - No redundancy (examples – gas turbine engine, generator, load gear).

Secondary Plant servicing the Unit where failure of the equipment would mean loss of the generating unit. – Full redundancy, e.g. 2 x 100%, or 3x50% capacity. (Examples – lube oil pumps, cooling systems, fans).

Secondary Plant servicing the Unit where failure of the equipment results in exceeding/non-compliance with environmental or operating licence conditions or significant risk to the safety and health of personnel or damage to equipment – full redundancy.

Secondary Plant servicing the Unit where failure of the equipment does not result in the loss of the generating unit – no redundancy.

Critical instrumentation where failure of the equipment would mean loss of the generating unit, exceeding the environmental or operating licence conditions or significant risk to equipment or personnel – full redundancy.

15.5 Permits Approval and Owner Insurance

The Engineering team will provide the necessary engineering data, drawings and calculations to support the permitting process as defined in the Project Permitting Plan and Prime Contract < make specific reference to such a plan or sections of the Contract if available>

The following is an indicative list of engineering documents required:

- Building plans and specifications for flammable storage or processing buildings, rectifiers and electrical rooms, and control rooms.

- Fire detection and alarm system specifications and design drawings.

- General arrangements for flammable/combustible liquid storage.

- Fire water distribution and installation drawings.

- Design calculations and drawings of fire extinguishing system such as fire hydrants, sprinklers and deluge systems, portable extinguishers, and gas-based fire quenching and any other dry chemical systems.

- Architectural design drawings for process and non-process buildings.

16
PROPOSED CONTENT FOR THE "ENGINEERING DESIGN MANAGEMENT" SECTION

A sample write up of this section can be as follows:

16.1 Management and Control

The design output will include calculations, analysis, specifications, data sheets, reports, feasibility studies and drawings as appropriate and in accordance with the requirements of the Contract.

The relevant Design Engineer/s preparing the document will ensure the following aspects of the design output:

Numbered, dated, indexed, and revision controlled.

Complies with design input requirements including the current revision of any vendor data.

Contains or references acceptance criteria, where appropriate, by quoting relevant codes, design loads, etc.

Conforms to appropriate design standards and regulatory requirements whether these have been stated in the design input.

Incorporates those characteristics of the design that are crucial to the safe and proper functioning of the works.

Each document is verified (checked) independently of the originator. When the originator has incorporated comments, the document is reviewed by the relevant Lead Discipline Engineer in the project team and approved by the Engineering Manager, or his designate.

When conflict occurs during the verification process the issues are discussed with and resolved by the Engineering Manager and the Lead Discipline Engineers.

Documents issued prior to "issue for construction" or "released" are generally stamped preliminary or noted in the revision description and are not necessarily approved.

Approval of the Project Manager may be required in cases where preliminary drawings are required to be released for construction purposes.

The document number and a listing of all issues, revisions, additions and deletions of individual document sheets are recorded on the document front sheet.

Document Control manages all aspects related to the control and issuance of these documents.

16. 2 Design Preparation

16.2.1 Issued for Design

Engineering documents/drawings prepared for design definition purposes will be notated "Issued for Design" when approved for release by the Engineering Manager or his designate.

These drawings will then be issued to the <Client> for their assessment and comments.

16.2.2 Approved for Design

Following receipt of direct approval and/or the incorporation of comments from the client, these documents and/or drawings will be up-revved to "Approved for Design" purposes and used for design development purposes.

Where no response is received within the stipulated period of twenty (20) working days from the client, the documents and/or drawings may be automatically up-revved to the "Approved for Design" status based on assessed risk.

16.2.3 Issued for Tender

Documents or drawings prepared for tendering purposes will be notated at "Issued for Tender" when approved for release by the Engineering Manager or his designate.

These documents and/or drawings will then be issued to the <Client> for a review period of twenty (20) working days for assessment.

16.2.4 Approved for Tender

Following receipt of direct approval or the incorporation of comments from the client, these documents and/or drawings will be up-rated to "Approved for Tender" purposes and incorporated in tender documentation.

Similarly, where no response is received within the stipulated period of twenty (20) working days the documents and/or drawings may be automatically up-revved to the "Approved for Tender" status based on assessed risk.

16.2.5 Issued for Construction

Documents or drawings prepared for construction purposes will be notated as "Issued for Construction" when approved for release by the Engineering Manager or his designate.

These documents and/or drawings will then be issued to the <Client> for a review period of twenty (20) working days for assessment.

16.2.6 Approved for Construction

Following receipt of direct approval or the incorporation of comments, these documents and/or drawings will be up-revved to "Approved for Construction" purposes and issued to the Site Construction Manager and the relevant contractors.

Similarly, where no response is received within the stipulated period of ten (10) working days the documents and/or drawings may be automatically up-revved to the "Approved for Construction" status.

16.2.6 Management of Red line Marked Up Drawings

During construction and commissioning, drawings will be marked up on site using a red lined mark-up process to ensure all as constructed detail is retained.

Two copies shall be marked up, one copy to remain on site and second copy issued to the Drawing Office via Site Document Control for update.

16.2.7 As Constructed Drawings

Updated drawings to be checked against site red lines to ensure completeness of updates. At conclusion of the Project all required drawings will be revised to "As Constructed" status with all site mark-up and corrections incorporated.

These corrected drawings may be issued to the <Client> for review and approval before final issue.

16.3 Design Coordination and Interfaces

16.3.1 Design Coordination with Contractors

The design coordination with these contractors shall be guided by an Interface Management Plan between <Consultant Name> and the contractors.

This plan shall be developed by the Engineering Manager and/or his designate, which highlights the key deliverables and inter-dependencies of these deliverables to the overall engineering design output and schedule.

Such a plan will ensure a high degree of planning and scheduling coordination between the various contractors.

It will primarily contain a distribution of responsibility matrix and communication plan which will address issues such as: battery limits, interfaces and tie-ins; construction schedule, delivery of major mechanical equipment; commissioning schedule; and standards.

It is envisaged that the oversight and management of the Interface Management Plan will be the responsibility of a working group composed of the Engineering Manager, Principal & Lead Discipline Design Engineers, Area Managers, Construction Manager, and the applicable counterparts from the contractors.

Client representatives may be invited from time to time into this working group.

It is expected that the working group will convene, at least weekly during the initial stages and more frequently if required depending on the issues encountered during the design delivery process.

16.3.2 Vendor Design Interface

Vendor design will be reviewed prior to fabrication or shipment.

Various witness points and hold points will be established contractually with each vendor to ensure appropriate engineering approval is given to the vendor at each critical stage of manufacture and supply of equipment.

Particular emphasis will be placed on vendor data that is needed to complete the overall project engineering design. This includes equipment physical characteristics, such as dimensions, shapes, loads, and bolt down details, as well as electrical or control system connectivity details.

Vendors performing detail design will use the relevant <Consultant Name> Project standards, specifications, and design criteria, which will be specified and made a part of the vendors' contractual obligations.

16.3.3 Internal Interdisciplinary Design Review (IDR or Squad Check)

Design reviews shall be carried out as per the engineering schedule and will be conducted to coordinate and incorporate comments and requirements from the various engineering disciplines and across other work share offices (WSO) consultants and third parties, as required.

An IDR division of responsibility matrix as provided in **Appendix C** will be used to delineate when the responsibility is transferred.

Further an Engineering Review Quality Plan as provided in **Appendix D** shall guide the overall coordination and review of the detail engineering deliverables and drawings by design area, including those documents requiring review and approval by the client.

A squad check distribution matrix will be produced to ensure all disciplines are covered as appropriate to each discipline's requirement.

Distribution of squad check documentation shall be via Document Control.
The squad check originator passes the document to Document Control with a completed document control work request form stating who checks the document, using the squad check distribution matrix as a guide, the timing and sequence of checking.

16.3.3.1 Physical Interface Tie-In Logs

All physical interfaces (tie-ins) on the project will be shown (as deliverables) and marked with a tie-in identification number.

A log of the tie-ins will be kept by a designated engineer, asset manager, discipline engineer and area engineer and released to construction at the appropriate time for maintenance in the field.

The construction manager will be responsible for coordination with the plant operators to schedule and implement tie-ins during construction.

16.4 Constructability Reviews

Constructability reviews will be part of the design work process and these includes the reviews of engineering sub-contractors and WSOs.

Engineering and construction will identify deliverables for constructability reviews before those deliverables are issued for Construction (IFC).

Where a facility is designed in a three-dimensional (3D) environment, construction reviews will be done using 3D walkthroughs; otherwise two-dimensional (2D) drawings will be used.

The Engineering Manager or his designate will conduct the review sessions and record the observations and comments of the review team.

Each review comment will be resolved by engineering and the actions taken will be recorded.

16. 5 Certification and Stamping

The design drawings and documents will carry the certifications and stamps in accordance with regulatory requirements.

The engineers and architects (as appropriate) shall be registered in the client's project jurisdiction, as governed by local statutory requirements.

16.6 Quality Control

Quality control will be re-enforced by the addition of a senior engineer (QA Engineer) responsible for complete and consistent application of the relevant QC procedures and the full integration of the processes with the WSO and engineering consultants.

The QC activities shall be carried out in accordance with the Engineering Review Quality Plan as provided in **Appendix D.**

This plan shall be managed by the QA Engineer under the responsibility of the Engineering Manager.

The engineering requirement of the project shall be conducted using the following quality control methods:

Usage of experienced engineers and designers including the QA Engineer who are qualified and assessed to be competent by the Engineering Manager to perform their required duties.

Perform the engineering activities related to calculations, specifications, drawings, and materials requisitions according to established procedures and the Progressive Engineering Quality Plan.

Adherence to strict inter-discipline co-ordination of all drawings and documents and independent review and checking of all drawings and calculations.

As a minimum, prior to issue for construction (IFC), all drawings will be reviewed and signed-off by a Drafting Checker, Lead Discipline Engineer and Engineering Manager and Area Manager.

16.7 Quality Assurance

Quality assurance is a shared function which is led by the QA Engineer in collaboration with the Lead Discipline Engineers and the Engineering Manager.

The Lead Discipline Engineers and/ or their designated technical specialists will perform the reviews of key design drawings and documents in accordance with the Progressive Engineering Quality Plan and related engineering procedures.

These typically consist of discipline design criteria, process flow diagrams, general arrangements, site plans, P&IDs, and major equipment specifications and data sheets for a selected area or system.

Periodic functional reviews will be conducted by the Project Quality Assurance Manager in collaboration with the Project Manager to verify that the work is being performed in accordance with the requirements of the Progressive Engineering Quality Plan and related engineering procedures.

16.8 Change Management

Every change from the baseline scope of the project, having an impact on technical content, design criteria, scope, cost and/or schedule will require early communication to the Client, Project Manager and other relevant parties for resolution and approval.

The change management process is managed by the Project Controls group which is responsible for the preparation and management of the change management procedure and the trend procedure as found in that group's list of procedures **(Appendix A)**.

Section 24 of this document provides detail description of the change management process that is adopted in this project.

The engineering team, led by the Engineering Manager will identify, define, and document potential engineering trends as they occur and actively participate in the approval/resolution process.

The project controls team is responsible for input into the TREND register and Change Notice register, which will be reviewed on a weekly basis by both the Engineering Manager and Project Manager to assure that there are no surprises in the project scope and/or cost.

Further details on trend management can be found in the Project Controls section of the Project Execution Plan.

16.9 Documentation Review and Approval by <Client>

The Contract has nominated review points namely at 15%, 50%, and 80% of design progress.

Refer to **Appendix L** for list of drawings and documentation that are to be issued for approval.
All such documents will be marked up on the Deliverables Register and will be issued to the <client> for review.

Design documentations submitted for review during the performance of the works will be in electronic format.

In addition, <client>. will be required to participate in other reviews such as HAZOPs, HAZIDs, LOPAs and design safety reviews as required during the design process.

<Client> has 20 business days to review documentation submitted as per the contractual agreement.

The Design Documentation Review Process involves the following 9 progressive stages as listed below:

1) Issue for 15% Internal Design Review;
2) Incorporate Comments and Review with <Client>;
3) Incorporate <Client's> Comments
4) Issue for 50% and 85% Internal Design Review
5) Incorporate Comments and Review with <Client>
6) Incorporate <Client's> Comments & Finalise Design
7) Issue for Internal Approval;
8) Issue for Client Approval; and
9) Issue for Construction.

16.10 Design Verification and Validation

Verification and Validation are independent processes that are used together for checking that a product, service, or system meets requirements and specifications and that it fulfils its intended purpose.

Validation can be expressed by the query "Are you building the right thing?" and verification by "Are you building it right?"

Verification can be defined in a little more detail as "The evaluation of whether or not a product, service, or system complies with a regulation, requirement, specification, or imposed condition".

Typical 'inputs' for the verification process are requirements and design representations such as 2D/3D mechanical CAD drawings and models, electrical schematics, and software code.

Typical 'outputs' are a determination of whether the design component or system met requirements, descriptions of failure modes, summarized test results, and recommendations for design improvement.

All design outputs will be subjected to formal technical review (checking) and approval by authorized personnel according to the design control procedure <specify procedure doc number if available>.

In the case of verification of calculations, the person performing the check will, where practicable, use at least one alternative method to verify the correctness of the original analysis or calculation.

If an output is to be modified as a result of a review, or as a result of a project change control process, this modification will be subjected to the same constraints as those applying to the production of the original output before re-issue for checking.

Documents will carry an alpha revision indicator, i.e. A, B, C etc, until the document has been "Approved", "Issued for Construction", or "Released", at which stage the documents are indicated as revision 0 (zero). Any revision indicator thereafter is numeric, i.e. 1,2,3 etc.

Revision triangles are removed from a document when it is "Issued for Construction", "Approved", or "Released", i.e. all the alpha revisions nominations within the document content are removed.

The Alpha description(s) remain in the revision box, but the content of the document is cleared at the Rev 0 issue.

Changes made from one issue to the next are described in the revision box on the document, commencing on the bottom line and working upwards.

Only the Project Manager and Engineering Manager or their nominees, can approve project documents, generally after all reviews, performed by the internal, client or statutory parties have been completed.

Formal and documented design reviews will be conducted at predetermined points during the design phase of the project. These design reviews will focus on identifying hazard and operability issues, possible omissions and constructability aspects.

16.11 Drawing and Document Approval Process

The approval of all internal design documentation is to be identified at project initiation by the Engineering Manager and Principal Engineers.

No drawing is to be issued to the Client without the relevant internal checks and approvals completed as described in the related sections above.

The table below refers to the design review process that will be undertaken for each design package which will be arranged by the relevant design engineer/s.

Design Reviewing Matrix		
Reason for Issue	**Rev**	**Required Signatures**
Drawings for Internal Processing and Review	Alpha	• Author/Design Office • Checked • Reviewer • Approver
Preliminary Drawings	Alpha	• Author/Design Office • Checked • Reviewer • Approver

Issued for Approval	Alpha	• Author • Reviewer • Approver • Wet Signature Required
Issued for Construction (Drawings issued for Construction and form part of the Construction Package)	Numerical (Generally, 0+)	• Author • Reviewer • Approver • Wet Signature Required
Issued for Use (Usually issued to vendors for guide or part of tender packages)	Numerical (0+)	• Author • Reviewer • Approver • Wet Signature Required
As Built	Numerical (Generally, 1+)	• Author • Reviewer • Approver • Wet Signature Required

The Engineering drawing office will ensure CAD Conform process is used on all drawings submitted to <client> for review and approval.

All drawings shall to be prepared and submitted progressively in work packs (rather than all at once).

The design documentation shall be issued in the relevant format via Project Document Control to the Client.

The <Client/Consultant Name> Drawing procedure (Specify Document Number – Documentation Requirements) shall be adopted closely in meeting <Client Name> requirements."

17
PROPOSED CONTENT FOR THE "HEALTH, SAFETY AND ENVIORNMENTAL (HSE) MANAGEMENT IN DESIGN" SECTION

A sample write up of this section can be as follows:

17.1 General

During the performance of the Works, engineering activities shall be cognisant of all aspects of health and safety (H&S) requirements of the Contract.

All engineering personnel and engineering work activity will comply with the requirements of the H&S Management Plan (HSMP) <specify Plan document reference number if it's available> and statutory requirements at all times.

The H&S management systems, policies, standards, procedures and work instructions will be progressively developed in consultation with <client>.

17.2 Environmental Requirements

All engineering for the Works are completed in full compliance with the Environmental Management Plan (EMP) <specify Plan document reference number if it's available> as well any requirements stipulated by <client> and relevant Laws.

<Consultant Name> will assist in providing <client> with the relevant information required to obtain environmental permits.

17.3 Safety in Design

The 3D Model Design Review software automation will be used for on-Project, HSE Reviews (Safety by Design) and design reviews.

Engineering and design work shall adopt best practice principles in design which address the following points as a minimum requirement for construction, commissioning and operations of the Works:

Eliminate identified hazards or reduce attendant risks through design, including material selection or substitution. When potentially hazardous materials must be used, select those with minimum risk throughout the life cycle of the facility;

Ensure adequacy of structure during construction activities;

Ensure adequate attention is given at the design stage to safety of personnel during construction and O&M;

- Locate equipment so that access during construction and O&M minimizes personnel exposure to hazards;
- Minimise risk arising from ambient conditions including noise and vibration;
- Design to minimize risk created by human error in the operation and support of equipment, facility or system;

- Design to minimize risk created by human error in the operation and support of equipment, facility or system;
- Consider alternative approaches to minimize or eliminate risks;
- Protect critical systems by shielding or physical separation from other hazardous equipment;
- Minimise the severity of personnel injury or damage to equipment in the event of mishap; and
- Review design criteria for inadequate or overly restrictive requirements regarding operability and safety.

Under the legal framework in <specify country/region>, at the end of the project <Consultants Name> are required to hand over a "design decision log" to <Client Name> and hold this as a record of what solutions were considered during engineering and why a particular engineering solution was implemented.

An Engineering Management Register shall be developed to record such key decisions. This register will maintain issues from formal design reviews and internal engineering meetings and client meetings.

Part of this requirement will be covered by formal design reviews where key decisions will be recorded.

This register will then be reviewed regularly by the Engineering Manager and Principal Engineers to ensure information is captured, managed and resolved effectively in a timely manner.

Once deliverables are issued for design then this process will be managed through the engineering change management process (refer to section 16.8).

The following key principles of safety in design will be considered though the engineering lifecycle during the detail design:

- Simplicity of design;
- Fit for purpose;
- Safety philosophy;
- Fire protection;
- Evacuation, escape and rescue;
- Working environment;
- Construction installation philosophy and interfaces;
- Storage and handling of chemical and dangerous goods;
- Bunding and disposal of liquid wastes;

- *Criticality and complexity of equipment;*
- *Space availability;*
- *Equipment access for maintenance and operation; and,*
- *Emissions such as noise, dust, flare, etc.*

These aspects will be considered as part of the model reviews, IDR and basis of design as applicable."

18
PROPOSED CONTENT FOR THE "RISK MANAGEMENT" SECTION

A sample write up for this section can be as follows:

"An Engineering Risk and Opportunity Assessment Workshop shall be planned at the onset of the project which focusses on the following:

- Risk Identification;
- Risk Mitigation Strategy;
- Management of risk registers; and
- Review, close out and reporting process.

18.1 Objective

The primary objective of this Workshop is to identify, describe, analyse, quantify and evaluate risks and opportunities associated with all areas of the Engineering activities including their impact on facility performance, quality, schedule, procurement, logistics, construction, operations, project budget, health, safety, environment, security and general project performance and to finalise mitigation strategies in order to maximise opportunities and minimise the potential impact of any identified risk.

Strategies to implement, track, manage and report adopted mitigations in respect of the identified risks and opportunities will be implemented as required.

This will include:

the assignment of personnel to the ownership of each identified risk and opportunity;
to determine time and budget constraints for the implementation of risk mitigation and opportunity strategies; and;
to provide progress reports to project management.

The Engineering Manager in consultation with the Project Manager will define distribution requirements for this plan depending upon the participants involved in the Engineering Risk and Opportunity Assessment Workshop

18.2 Risk Management Approach

The overall risk management approach follows the standard risk management model of identify, analyse, plan, monitor/control as used in most risk management plans.

The following risk matrix shall be used in this project.

<incorporate a client approved Risk Classification Matrix e.g.4x4 or 5x5 matrix below or include this into the Appendix>

18.3 Design Risk Review

At the commencement of the design, engineering risk reviews will be carried out.

Each design package is analysed to determine whether there are competencies to design and/or review them in-house or should be designed/ reviewed externally.

Further matters to be considered during this review include:
- Application of new technology;

- Application of new construction techniques;
- Design complexity - specialist resource requirements;
- Fabrication complexity;
- Application of standard industry practices;
- Statutory requirements;
- Cost implications of failure and to project costs and schedule – critical path & lead time concerns;
- Personnel safety, both during construction and operation of facility;
- Hazards to the environment;
- Any special considerations or requirements;
- Project schedule risks; and
- Control of design output from Work share offices (WSO);

The intention of the review is to highlight potential problem areas that either requires resolution, additional verification or special action.

The following outcomes will be considered for each potential problem area:

- Issue of a specific technical instruction to discipline engineers or by updating the discipline engineering basis of design.
- Issue of a project specific engineering standard or procedure.
- Use of internal consultancy service, e.g. materials specialists, engineering specialists.
- Use of expertise from other <Consultant Name> offices through work sharing agreements.
- Reliance (or otherwise) on suppliers or subcontractors expertise.
- Use of client's or vendor's expertise.
- Use of an external consultant.
- Use of specialist engineering Inspection service.
- Use of alternative calculations/industry standards.

All key design work packages will be reviewed and analysed in this way and the outcome of this analysis is tabulated below, especially in relation to available in-house competencies.

<Below is a sample of the outcome of the high-level risk review carried out on the design work packages on the project>

Engineering Work Package	Interface Complexity	Design Skill	Cost of Failure	Risk Assessment	Engineering
Plant layout	Complex	Available in house	Moderate	Medium	As this task directly affects the project outcomes and <Client/Consultant Name> reputation it is
Protection Scheme	Complex	Available in house	High	High	An external design review is required.
Gas Turbine and Steam Turbine Main Foundations	Complex	Not available in-house	Extreme	High	Design must be done by a consultancy with proven experience in this field
Structural Supports for pipe and cable racks	Complex	Not available in-house	High	High	Design must be done by certified structural designers
Minor foundations	Moderate	Available in house	Low	Low	As we have limited in-house capability the designs must be verified externally or done externally
Main Steam Pipe Design	Very Complex	Not available in-house	Extreme	High	Design must be done by a consultancy with proven experience in this field
Water, condensate and other interconnecting pipework	Moderate	Available in house	Low	Low	Should be done internally
Electrical, earthing, lightning protection	Moderate	Available in house	High	Medium	Should be done internally

BOP instrumentation and control	Complex	Available in house	Moderate	Medium	Should be done internally
Procurement of Engineered Packages	Complex	Available in house	High	High	Should be done internally with a mandatory internal or external verification
Cable and Termination Schedule	Complex	Available in house	High	High	Should be done internally with a mandatory internal or external verification
BOP Plant Specifications	Complex	Available in house	High	High	Should be done internally with a mandatory internal or external verification

The Engineering Manager shall follow-up and monitor implementation of solutions to problems highlighted.

19
PROPOSED CONTENT FOR THE "CONSTRUCTABILITY REVIEWS" SECTION

A sample write up for this section can be as follows:

"Constructability is defined as the systematic use of construction knowledge and experience in planning, design, procurement, and field operations to achieve all project objectives.

Constructability objectives include primarily the following:

- *Ensuring work safety, quality and environmental requirements;*

- Reducing total installation costs;
- Improving schedules;
- Facilitating pre-commissioning, commissioning and operation.

Constructability reviews will be held during the preliminary design and detail design phase.

The constructability review will use the 3D model as a tool to conduct the evaluation and outcomes will be formally recorded and actioned similar to the design risk review process.

Participants for these reviews will include Construction Manager, Discipline Construction Supervisors, Engineering Manager, <client representatives>, Principal Engineers, Lead Discipline Engineers and Safety/Environmental Manager as required.

Key documents that would be reviewed at these meetings include the following:

- equipment lists,
- site area maps,
- site plot plans,
- layout drawings,
- detail drawings, and
- project schedule.

Key discussion items include the following among others:

- Health, safety, and environment (HSE);
- Logistics;
- Module/Preassembly fabrication sites organisation, facilities, access and management systems;
- Infrastructure and site access;
- Heavy equipment and vehicle parking areas;
- Temporary power;
- Temporary Communications;
- Bulk storage, warehouse and laydown areas;
- Flood mitigation and contingency planning;
- Water basin excavations, spoil storage and disposal;
- Plant site access and site preparation;
- Procurement, delivery and installation schedule;
- Crane and personnel access;
- Foundations locations;
- Underground services;
- Overhead obstructions;

- Crane-lifting capacities to loads.

The review committee will generate a report and action list with assigned responsible individuals.

The clearing of engineering action items by agreed dates will be the direct responsibility of the Engineering Manager and the relevant Principal/Lead Discipline Engineers.

20
PROPOSED CONTENT FOR THE "MAINTAINABILITY AND OPERABILITY REVIEWS" SECTION

A sample write up for this section can be as follows:

The maintainability and operability review will also be conducted as part of 3D model reviews.

Key documents that would be reviewed at these meetings are:

- *layout drawings,*
- *vendor drawings (GA's);*
- *any available maintenance manuals and*
- *the 3D plant model.*

Participants will mainly include <Client maintenance and operations personnel>, Area Managers, Engineering Manager, Principal Engineers, Lead Discipline Engineers and HSE Manager (as required).

Key discussion items would include the following at minimum:

- *Health, safety, and environment, e.g. spillage of lubricants, chemicals and fuels;*
- *Access for maintenance personnel and their specialised equipment;*
- *Withdrawal space for removable components and sub-assemblies; e.g. heat exchanger bundles and turbine rotors;*
- *Pipe and cable routes;*
- *Crane lifting plans, operating rules and regulations;*
- *Control and equipment rooms inspection and maintenance access;*
- *Motor control centres inspection and maintenance access;*
- *Communications centre and tower;*
- *Workshop, administration and control room areas.*

This review will generate a report and action list with assigned responsible individuals.

The clearing of engineering action items by agreed dates will be the direct responsibility of the Engineering Manager and the relevant Principal/Discipline Lead Engineers.

Relevant action items will also be incorporated into the plant operations & maintenance manuals.

21
PROPOSED CONTENT FOR THE "MODULARIZATION AND PRE-ASSEMBLY" SECTION

sample write up for the section can be as follows:

The concept of offsite modularisation and pre-assembly will also ensure the maximum amount of work related to the package that can be carried out offsite, with a minimum amount of work required onsite for construction and integration into the facilities.

The final determination of offsite modularisation and pre-assembly will be dependent on the maximum dimensions (length, width, height, volume, weight), which can be achieved with the current access routes to the site as well as with the available transportation vessels/vehicles.

The appointed project team parties will carry out surveys of the access routes to the site and this information will be used to determine the maximum dimensions.

Items which may be included for the application of modularisation are:
- prefabricated buildings,
- parts of process plants mounted on a flatbed skid,
- parts of utilities areas mounted on flat bed skids,
- small tanks,
- pipe spooling,
- structural fabrication, etc.

21. 1 Construction Work packs

The Engineering team will prepare discipline specific construction work packs during the detail engineering for the works.

Where practicable, the work package will represent one discipline only, be it piling, civil, building, structural, piping, mechanical or electrical/instrumentation etc.

The work packages will be:

- incorporated as identifiable elements within the construction schedule and progress will be monitored at this level.

- numbered in accordance with project numbering procedures and will be collated and issued via document control.

They will be sufficient to cover construction requirements and will typically contain the following documents and information, as applicable:

- Work scope.
- List of drawings applicable to the scope.
- Vendor data and equipment installation instructions.
- Special instructions, including references to specifications, NDT and heat treatment requirements.
- Materials list and loose items list including availability.
- Lift studies as applicable.
- HSE requirements.
- Completions data (field inspection check sheets, pre and post installation).

These sheets will contain as a minimum:

- Fabrication check sheets.
- Punch list and handover check sheets for pre-commissioning.
- Separate test packs will be compiled for hydro testing.

21.2 Preparation of Pre-commissioning System Work Packages

The early definition of plant systems within and across facilities during the detail design phase as well as preparation of the mechanical completion work packages as well as system completions work packages will assist in the preparation for pre-commissioning activities for the works.

Each of these work packages will be assembled during the engineering phase and be made up of documentation from the respective disciplines and will also include any specialist procedures such as air blowing, vendor attendance etc.

Part of the preparation of the system work packs will take into account the possibility of the system passing through different construction areas. The system definition and completion scheduling will make certain that construction activities in these areas will support the achievement of the pre-commissioning activities as an integral part of the works assembly.

A completions database <specify type of database, if available> will be utilised to generate the required pre-commissioning check sheets and checklists.

The pre-commissioning work pack numbering will be consistent with construction work pack numbering and will be defined by the project numbering procedure.

Further details of the requirements under this section will be expanded in the Commissioning Plan which will be developed and issued in due course.

21.3 Engineering Support during Construction

The Engineering team will provide follow up engineering and technical support during the assembly phase. The team will provide engineering resources for clarification of drawings and documentation, and where required, engineering follow up of the assembly work, pre-commissioning assistance.

A site engineering team will be established which will work with the project office design team prior to locating to site.

All changes and/or improvisations to the IFC drawings due to site considerations shall be basically governed by the management of change process.
Authorization of all changes/modifications shall require the review and approval of the Engineering Manager and/or Area Project Managers or their delegates.

Where these changes/modifications are significant and may result in considerable schedule and cost impact, these will need the approvals of the Project Manager and Client.

Refer to Section 24 of this document for the Change Management Process.

22
PROPOSED CONTENT FOR THE "ENGINEERING DRAWING MANAGEMENT" SECTION

A sample write up for this section can be as follows:

22.1 Engineering Drawing System

Engineering uses <specify the system used> to manage the drafting, review and coordination of all designs prepared for this project.

Once the designs are ready to be issued, this will be issued via the Project Document Controller.

The overall drawing management process shall be governed by the procedure <specify procedure doc reference, if it's available>.

22.2 <Client Name> Drawing Requirements

Requests for <Client Name> drawing numbers and templates for drawings will be submitted via Project Document Control in accordance with <Client Name> drawing numbering requirements procedure.

The <Client Name> standard for drawings is <specify> Format. Submission of AutoCAD for any submission can only be supplied for "off the shelf" products, all other drawings will need to be in <specify> format.

The following <Client Name> drawings and documentation procedures for conformance requirements shall be referenced:

- <Doc Ref No > Rev 0 – Drawing Specification.
- <Doc Ref No > Rev 0 – Drawing Procedure.
- <Doc Ref No > Rev 0 – Documentation Requirement Procedure.

Preliminary drawings (alpha revision) will be submitted by issuing PDFs.

Issued for construction (IFC) drawings (numerical revisions from Rev 0) will be submitted by Issuing PDF of the wet signature drawing (1 x A3 wet signature, PDF & DGN on CD).

As Built (numerical revision) will be submitted by issuing PDF of the wet signature drawing and CAD file (1 x A3 wet signature, PDF & DGN on CD).

22.3 Design Approval Coordination

On drafting completion, drawings are circulated for checking and approval via squad checking process through Project Document Control.

All amendments by the various review engineers will be reflected via a colour coding system. Drawings are then returned to the designer to be either changed or back CAD drafted for enlisting internal approval.

Once drawings are approved and signed off internally, the designer back drafts all the signatures, and binds all external references to freeze that revision and then purges the drawing and creates a PDF and is marked up "issued".

Drawings are then processed through CAD conform prior to being submitted to the client for approval. This will ensure that the drawing conforms to the client's CAD standards and avoid being rejected for non - conformance.
All revisions to drawings shall be written to the CAD file. Each revision shall be indicated by revision triangle cell with the revision number drawn within.

The drawing will then be submitted for review to the Client and is either returned with comments or issued as "approved for construction". Drawing revisions that are "approved for construction" are issued under numerical Rev 0, 1, 2.

At completion of the construction phase the constructed mark ups are returned to the design office and are up revved to the next consecutive numerical revision.

The drawings are modified as per the mark ups and then processed through CAD Conform and submitted for both internal and client for sign off.

All designs are to be documented at all stages as checked, recorded and filed for future reference.

22.4 Design Mark Ups

Design development and mark ups will be managed by the Design Office using <specify system & tools used>.

22.5 Cancelled Drawings

This applies to drawings where the information is no longer valid.

This requirement does not apply to normal drawing revisions.

No cancellation of a historical revision shall be done because the next revision of the same drawing has superseded it.

The comments field shall record details of the cancellation to maintain traceability of drawing information with the person's initials.

This drawing is then returned to the originator who will mark up the drawing and register information on the drawing register.

22. 6 Numbers

If drawings have been received and registered by Drawing/Design Office and are subsequently cancelled, the drawing number shall NOT be used again.

For traceability, the drawing to be cancelled shall be revved-up, a brief statement made in the revision block and diagonally across the drawing in large text between two parallel lines the wording.

22.7 Red Line Marked Up Drawings

During construction and commissioning, drawings will be marked up on site using a red lined mark-up process to ensure all as constructed detail is retained.

Two copies shall be marked up, where one copy remains on site and the second copy is issued to the Drawing Office for incorporating updates.

Updated drawings shall be checked against red lines to ensure completeness of updates.
At the conclusion of the project all required drawings will be revised to "as constructed" status with all site mark-up and corrections incorporated.

These corrected drawings will be issued to <client> for review and approval before final issue via Project Document Controller.

On completion all engineering drawings, documents, data, quality assurance and quality control requirements, inspections, testing and certification requirements will be provided to <client> in accordance with the requirements of the Contract Terms and Conditions.

All drawings and documents shall comply with the requirements of < Client Name>

- <Doc Ref No> Drawing Procedures; and
- <Doc Ref No> Document Requirements

23
PROPOSED CONTENT FOR THE "ENGINEERING COORDINATION WITH OTHER FUNCTIONS" SECTION

A sample write up for this section can be as follows:

"23.1 Coordination with Procurement
23.1.1 Establishment of A Procurement Plan

Engineering will work together with Procurement to develop the procurement plan for procuring of major equipment.

This Plan is developed and maintained by the Procurement Manager with regular inputs from the engineering team.

This plan will identify items to be procured, how they will be procured, identified suppliers, the inspection levels, how items will be packaged for market, the tendering process that is to be adopted, technical criteria for selection, tender assessment for compliance and how the contract will be managed once executed.

23.1.2 Procurement Deliverables

The Engineering team will produce procurement checklist for all major equipment which is to be procured.

This checklist outlines equipment item numbers, design methodology to be used, technical specification for tendering purposes, suppliers/tender information if available and information which is required to be supplied.

<Refer to an Engineering Procurement Checklist Template if available>

The Lead Discipline Engineers are responsible for ensuring these checklists are developed on time, assisting the procurement team with technical queries and tender clarifications as well as tender reviews, during the tendering phase.

Reference shall be made to the following procurement documentation:
- Procurement Plan; and
- Procurement Status & Expediting Register.

23.1.3 Procurement Deliverables

The objective of the Engineering team is to ensure that the tender responses are technically evaluated in accordance with the requirements specified in the tender documentation.

23.1.3.1. Evaluation Team

The engineering tender evaluation team shall comprise at a minimum of two (2) design engineers and a Lead Discipline Engineer.

They will normally be the Lead Discipline Engineer as named in the tender documents for engineering queries plus the other two design engineers.

A member of the Procurement team shall be available as required.

23.1.3.2 Evaluation Criteria

The evaluation criteria adopted shall be based on the technical specification in the tender documents as well as the following considerations:

- Completeness of response.
- Conforming bid.
- Alternative or non-conforming bid(s) clearly defined.
- Addresses all technical requirements.
- Provides an accurate and complete list of non-conformances.
- Bid does not contain conflicting or contradictory technical information.
- Provides clarifying comments on non-conformances.
- Contains appropriate and sufficient level of technical detail.
- Includes sufficient detail in respect of factory testing and sign off of equipment.
- Includes full details in respect of packing and delivery methodology.
- Experience and professional and technical competence.
- Understanding and ability to comply with required external technical standards (e.g.<CLIENT/CONSULTANT NAME> specific standards).
- Capability / Capacity.
- Experience in the Industry – List of past/present contracts undertaken by the tenderer's organisation supplying similar equipment.
- Specific relevant past performance.
- The tenderer's personnel, qualifications and experience who are involved in the equipment design, manufacture, test packing transport and delivery.
- Ability to respond in a timely manner to RFIs or technical presentations.
- QA processes and procedures that will be applied.
- Compliance with project schedule.
- Preparation for transport, delivery methodology and risk.

23.1.3.3 Evaluation Process

Each evaluation will be conducted in accordance with an agreed evaluation plan and matrix.

This plan will nominate the team members. The level and content of the plan and team makeup will depend on the size, complexity and risk in respect of the equipment to be procured.

The evaluation plans and associated matrix will be signed off by the Engineering Manager, Contracts & Procurement Manager and the Project Manager.

It should be noted however that some standard low cost/risk equipment will be directly purchased and will therefore not be based on a tender process.

Prior to completion of an evaluation report and evaluation matrix, it may be necessary to conduct interviews or request presentations by tendering organisations.
Two or more persons from the evaluation team should be present at any such meeting.

All presentation material and the minutes of any meeting will be included as part of the evaluation package.

23.1.3.4 Evaluation Report and Recommendations

Each technical tender evaluation and recommendation will be prepared and signed off by the by the evaluation team.

It will then be reviewed and signed off by the Engineering Manager and released to the Contracts & Procurement Manager to facilitate the purchasing process.

A copy of the Technical Evaluation report and recommendation shall be forwarded to the <client name> for his review and approval prior to initiating the purchasing process.

Engineering will work closely with the procurement team and be co-located at the project head office, providing technical input for enquiries.

Engineering will develop the scope of work, determine the vendor data requirements and develop quality requirements in collaboration with the Quality Manager.

During the bid period, engineering will provide technical input to vendor queries in collaboration with the Contracts & Procurement Manager.

When bids are received engineering will perform technical bid evaluations by evaluating vendor bid compliance with the technical requirements.

Engineering will prepare all technical documentation required for equipment and material RFQs and purchase orders (POs).

These documentations will typically include:
- Scope of Work;
- Scope of Work Schedule 1, Document and Data Submittal Requirements;

- Scope of Work Schedule 2, Applicable Documents List;
- Scope of Work Schedule 3 – Master Inspection and Test Plans MITP (if required);
- Scope of Work Schedule 4 – Key Dates. The key dates are nominated dates for issue of key documents to <Consultant Name> to allow the detail design process to progress efficiently;
- Quality Assurance requirements;
- MTO's;
- Technical Specifications;
- Technical datasheets;
- Drawings;
- Spare Parts requirements and lists.

The above technical elements of the RFQ / PO will be checked by the responsible Discipline Lead Engineer.

Upon the incorporation of all review comments onto the documents, the originating engineer will sign off as the Package Engineer and obtain approvals from the following parties:

- Lead Discipline Engineer;
- Contracts & Procurement Manager;
- Quality Assurance Manager;
- Engineering Manager;
- Commercial Manager; and
- Project Manager.

Vendor data requirements and delivery schedules will be clearly defined in each purchase order. Procurement and Document Control team will expedite vendor information to facilitate quick and efficient workflow during design in accordance with the project schedule.

As the design progresses, engineering will prepare material take-offs for bulk materials to allow timely purchasing to occur.

There will be an initial bulk material order based on preliminary quantity estimates, which will be updated during detail design usually as an intermediate and final update. Revisions to the material take-offs will be controlled and issued in accordance with project schedule requirements.

A weekly internal Procurement / Engineering coordination meeting will be held and minuted by Procurement.

These meetings will discuss the Procurement Status Register (PSR) and any issues with the development and approval of RFQ / PO packages.

The engineering attendees to these meetings will include the Engineering Manager and/or Lead Discipline Engineers.

Package engineers may be invited to discuss particular packages as necessary.

23.2 Coordination with Fabrication Assembly

The Quality Manager who is also responsible for the quality assurance and quality control of the fabrication scope of works as well as the Construction Manager who is responsible for the onsite assembly of the fabricated items, will be involved during the engineering and design activities in providing advice on fabrication and construction related issues.

This advice will include defining the requirements of the MDRs, ITPs, construction plant access, maximum crane requirement limits, potential for modularisation and definition of off-site fabrication.

During construction the Construction Manager will be assisted as necessary by site engineering group for routine technical queries and quality management initiatives supported by the Engineering team from the project office.

23.3 Coordination with Project Controls, HSE and Quality

23.3.1 Project Controls

Engineering will provide input to the project planner by way of progress reports.

Mark-ups of the project schedule based on the progress will also be issued to the project planner on a fortnightly basis for updating of the master schedule.

Engineering will also provide input into the various other weekly and monthly reports as required.

23.3.1 HSE

All engineering personnel will be involved in the communication and decision-making processes, as required especially on aspects that require design considerations and/or modifications to existing facilities etc; for the management of the Health, Safety and

Environment throughout the project in accordance with the Project Health & Safety and Environment Management Plan.

Some of the key coordination and guidance from the Engineering team with the HSE team involves the following:

- Equipment certifications and design verifications to Work Safe requirements;
- Finalisation of Hazardous Area Dossier;
- Construction related permits and licenses;
- Early works approvals and Environmental Management Plan (EMP); and
- Classified Equipment Registrations.

23.3.1 Quality

Engineering management audits will be conducted progressively to suit the project priorities and in accordance with the requirements of Project Audit Plan, which is finalized by the Quality Manager and agreed with the Project Manager.

Engineering will also review and approve all inspection and test plans (ITPs), as part of the co-ordination with Procurement and Quality Inspection team in the preparation and documentation of RFQ's and purchase orders.

Engineering will also determine criticality ratings for procured items, equipment, packaged units or services in accordance with the <Procedure for Determining Criticality of Engineered Equipment, Material and/or Packaged Units – XXXYY: incorporate such a procedure if available>.

The criticality ratings can be used to select the appropriate quality controls and the levels of inspections to be applied during the procurement process.

Engineering shall guide and coordinate closely with the Quality Inspections team, Site Quality team as well as the Package Engineers in ensuring that the procured items have had adequate off-site inspections in ensuring that they are dispatched from the vendors' premises without defects.

23.4 Design Office and Vendor Interface Management

23.4.1 Design Office Interface Management With WSOS

All Design documentation shall be located within one central repository located at <Consultant Name> Design office.

The Design office shall ensure at all times that Document Control has been provided copies of all relevant design documentations which requires to be controlled.

Other WSO design offices <specify who these parties are> will produce design documentations in accordance with their design deliverables list with respect to the packages they are assigned with.
Refer to section 11.6 which outlines the overall communications process flow.

23.4.2 Design Office Interface with Client

All formal communications to the Client shall be through letters or technical queries. These shall be channelled through the Project Manager and/or Engineering Manager respectively.

All correspondence and drawings issued to the Client around design issues and clarifications are forwarded via transmittal through Document Control so that the project has a central repository of all communications coming in and out to the Client.

23. 4. 3 Vendor Data Interface Management

All documentation received to and from Vendors shall be registered through Document Control. This information is captured and uploaded into <specify doc con software that is used> with notification sent to the Package Engineer of the receipt of such.

During the tender phase, Document Control will provide the relevant vendors with the start-up vendor packs for drawings, which specifies <Consultant Name> requirements for design documentation, procedures and templates.

The Package Engineer shall ensure that the vendor deliverables submission schedule meets the design reviews that need to occur especially around major procurement items and construction.

The < Client/Consultant Name> Document Control Procedure <specify doc number and title> shall provide the overall guidance in managing this interface."

24
PROPOSED CONTENT FOR THE "CHANGE MANAGEMENT" SECTION

sample write up in this section can be as follows:

24.1 *Engineering Change Requests (ECR)*

Critical design documents, such as Basis of Design, plot plan, P&IDs and electrical single line drawings etc. are controlled after being "issued for approval" at Rev 0.

These documents will then form the baseline design documentation.

Engineering Change Request (ECR) forms <specify document reference number if available> shall be used to monitor all design changes which occur after approval of such documents.

Details of the proposed change, its source, and the affected documents are recorded on the ECR.

All changes to design will be controlled and monitored by the engineering team.

Any change (including responses to TQ's and EN's), which will affect the baseline data in the Contract or in any way affect the scope, budget, quality or schedule, design, procurement, planning, scheduling, resourcing or project costs will be recorded and distributed through Document Control.

A thorough investigation will take place to investigate the extent of the change and the subsequent impact(s) it has to the project. No changes to the base design documentation will be made without an approved Engineering Change Request.

All ECRs will need to be authorized by the Area Manager and Engineering Manager on most occasions, unless by their assessment the ECRs are significant and need to obtain the approval of the Project Manager and/or the Client.

The Package Engineer in collaboration with the Engineering Manager is responsible to ensure that the Contracts and Procurement Manager is involved once the extent of the design change is known and whether a variation to the contract is warranted.

All significant change management issues are recorded in the project register, which are controlled, monitored and investigated with mitigations applied.

24.2 Request for Information / Technical Queries

Request for Information (RFI) and technical queries (TQ) may relate to the following:

- *Specifications;*
- *Technical Queries;*
- *Schedules;*
- *Scope of Works;*
- *Drawings / deliverables;*
- *Meeting minutes;*
- *Codes & Standards; and*
- *Regulatory requirements.*

24.3 RFI/TQs - <Consultant Name>/ <Client Name> Interface

All RFI/TQs issued to and by <Client/Consultant Name> will use the pro-forma referred to in **Appendix M** – RFI/TQ pro-forma.

To differentiate between RFI/TQs that are raised by various WSOs, originator centres sequential numbers will be allocated as follows
- Project Office   0001 - 0999
- WSO 1   1000 - 1999
- WSO 2   2000 – 2999
- WSO 3   3000 – 3999

Upon return of the RFI/TQ, the originator will determine whether the query has been answered satisfactorily or not.

If the query is not answered to the satisfaction of the originator, then a meeting is arranged with the interested parties to resolve the RFI/TQ.

Where there are cost/ construction/fabrication/schedule implications with the response to the RFI/TQ, it may result in the raising of an Engineering Change Request (ECR) and resolution thereof as detailed in Section 24.2.

24.3 RFI/TQs - <Client/Consultant Name> & Contractors Interface

All RFI/TQs issued by <<Client/Consultant Name> to the vendors/contractors and vice-versa will also use the pro-forma referred to in **Appendix M** – RFI/TQ pro-forma.

A numbering/document tracking convention shall be established in general for the RFIs/TQs which are to be raised by the vendors/contractors.

The Package Engineers shall be responsible for managing the RFI/TQ process related to their specific vendors/contractors.

They shall be required to engage with the Lead Discipline Engineers and/ or the Engineering Manager to resolve all queries.

In the event an ECR is required, the Package Engineers shall adopt the process outlined in Section 24.2."

25
PROPOSED CONTENT FOR THE "QUALITY MANAGEMENT" SECTION

sample write up in this section can be as follows:

25.1 Peer Review

As part of the Quality management process for the project, peer reviews will be conducted during the performance of the Works.

A nominated team by the Engineering Manager, comprising of Principal Engineers, Lead Discipline Engineers and Design Engineers shall form part of these peer reviews.

25.2 Lessons Learnt and Continuous Improvement

Upon completion of the project, a Lessons Learnt workshop shall be undertaken to evaluate and capture the key continuous improvement opportunities that future projects can benefit. Engineering will use a template approved by the Quality Manager to capture Lessons Learnt throughout the project duration.

25. 3 Records Management

All records will be uploaded and registered within the <specify the document management and control system> for transmitting, storage and version control.

Engineering shall maintain copies of all documentation within local engineering folders with a parallel folder mirrored onto the main project engineering folder.

Archiving and storage facilities are allocated which ensure that all stored/archived records are identifiable and retrievable.

Where records are maintained on computer magnetic media, these are subject to "back-up" at regular intervals, with the "back-up" information being stored in a protected location to ensure security from loss/ damage of active data.

On completion of the Project, all final drawings and documents will be copied across to <Consultant Name>'s filing structure in accordance with the Quality Management Plan and kept in archive for a minimum of 5 years.

Refer to the relevant sections of the Project Quality Management Plan <specify document number and title>"

26 PROPOSED CONTENT FOR THE "HANDOVER AND COMPLETIONS" SECTION

A sample write up for this section can be as follows:

26.1 As Built Drawings

Update of drawings for As Built will be undertaken as construction proceeds and accordingly the site engineers will complete redline mark-ups of the nominated documentation to be 'As Built' in accordance with the Contract.

The as-building of the drawings will be done as part of the engineering 'follow on' activities as each area of work is completed.

The <Consultant Name> Document Control Procedure <specify document number and title> shall provide the overall guidance in managing As Builts.

26.2 Vendor and Design Documentations

The handover of the vendor and design documentations shall be in accordance with the requirements of the <Consultant Name> Document Control Procedure <specify document number and title>.

The documentations are inclusive but not limited to the following:

- Basis of Design;
- Calculations;
- Technical Specifications;
- Reports;
- MDRs;
- IOMs;
- Installation Records; and
- Hazop Close Out Reports; etc.

26.3 Project Handover Data Books

Project data books, which include design documentation, vendor MDR's, vendor operating and maintenance manuals, high level plant operating instructions, QA / QC test records, site list and installation records, as-built drawings, engineering data and similar will be developed.

Compilation of this information will proceed in parallel with the development of the project so that this documentation will be available for reference before the start-up of the facility."

27 PROPOSED CONTENT FOR THE "APPENDICES" SECTION

Note: This is a proposed list only, it is scalable and editable, however what is shown is a sample list in a typical EMP

Appendix A	List of Applicable Engineering Procedures
Appendix B	List of Applicable Standards, Specifications and Acts.
Appendix C	Engineering Deliverables Matrix
Appendix D	Engineering Review Quality Plan
Appendix E	Engineering Deliverables Per Discipline
Appendix F	Engineering Organization Structure
Appendix G	Engineering Training Plan
Appendix H	Standard Automation Applications
Appendix I	Project Stakeholders and Engineering Team Directory

Appendix J *Engineering Schedule*
Appendix K *Engineering Quantity Management Flow Diagrams.*
Appendix L *Drawing and Document Schedule*
Appendix M *RFI/TQ Proforma*

www.ingramcontent.com/pod-product-compliance
Lightning Source LLC
Chambersburg PA
CBHW031430210526
45464CB00005B/2126